CALCULATIONS

A

PRACTICAL APPROACH

Kingsley **Anaba** BSc.,MSc.

Dedicated

To

My mother, Elizabeth

TABLE OF CONTENTS

- INTRODUCTION 4
- ATOM, MOLE & MOLECULE 6
- MOLECULAR AND EMPIRICAL FORMULA 17
- MASS AND VOLUME RELATIONSHIPS IN REACTIONS 28
- VOLUME RELATIONSHIPS FOR GASES 32
- VOLUMETRIC ANALYSIS 55
- DILUTION OF SOLUTIONS 61
- CALCULATIONS USING TITRATION RESULTS 65
- SOLUBILITY 102
- GAS LAWS 111
- BOYLE'S LAW 111
- CHARLES'S LAW 112
- OXIDATION NUMBER 132
- ELECTROLYSIS 138
- ENERGY AND CHEMICAL REACTIONS 157
- HESS'S LAW AND THE BORN-HABER CYCLE 157
- BOND ENERGY 161
- CHEMICAL THERMODYNAMICS 162
- RATE OF REACTIONS 165
- RATE LAWS 175
- HALF-LIFE CALCULATIONS 180
- EQUILIBRIUM CONSTANT 183
- SOLUBILITY PRODUCT CONSTANT 185
- HYDROGEN ION CONCENTRATION 194
- Practice Questions 204

INTRODUCTION

It can be quite easy to solve chemistry calculations if one has a basic or fundamental knowledge of chemistry. Generally, most questions in chemistry are like simple mathematical or aptitude questions. All that is required is for one to know how to get to the answer.

You must carry out an analysis of a question before attempting to solve it, look at what you are given and what you require to get your answer. Then think of how to get those values required to get the answer.

Having said that, it is worth noting the important fact that to be successful in solving calculations in chemistry, you need to know the basics. That is to say you need to read this book from the beginning all the way through. Just like medication you can't take what you like and leave the rest, you need to take everything in order for the treatment to work.

As you progress from the elementary to the more complicated stuff, the complicated stuff will make more sense. The reason I include this is that as a chemistry student I wasn't taught chemistry the easy way. Everything seemed so complicated until I started to look at everything on my own starting from the basic stuff all the way through. Then I realized that calculations in chemistry could actually be so simple for anyone to understand if it is presented in the right manner. Therefore, we are going to look at calculations in a practical manner. Understand the basic principles and then treat everything else with common sense.

ATOM, MOLE & MOLECULE

We are going to quickly review the meaning of some components which will pop up continuously as we deal with calculations in chemistry, before diving in. Below are some definitions we have given to some of those components to enable us understand the role they play in chemistry and how to measure and manipulate them.

Atoms are simply the building block of an element which posses the chemical properties of that element, Atoms contain particles called **Protons**, **Neutrons** and **Electrons**. When a group of atoms are bonded together to form a unit which can take part in a chemical reaction, we call them a **Molecule**. So molecules will have the chemical properties of the substance.

We measure chemical substances in chemistry using the unit, **Mole**. This makes it easy for us to understand the quantities of substances we are working with. 12 grams of Carbon -12 is used as our standard, so one **Mole** of any chemical substance contains as many particles as 12 grams of Carbon – 12.
How many particles does 12 grams of Carbon -12 contain?
It contains 6.02×10^{23} particles, so one Mole of any substance you will be working with will always contain 6.02×10^{23} particles. This is Avogadro's number.
Now we know how to measure the amount of chemical substances we are working with, we also need to be able to

weigh them. As in measuring we also get the weight by comparing them to Carbon – 12. So instead of saying the mass of one atom of a substance we say the Relative mass of the atom, or in other words the **Relative Atomic Mass**. The relative atomic mass of an element is the number of times the average mass of one atom of that element is heavier than one-twelfth ($^1/_{12}$) the mass of one atom of carbon-12.

The **Relative Molecular Mass** of an element or compound is the number of times the average mass of one molecule of it is heavier than one-twelfth the mass of one atom of carbon-12. Again to weigh the molecule of a chemical substance we compare it against Carbon -12.

The relative molecular mass of an element or compound is usually easily calculated by doing the sum of the relative atomic masses of all the atoms in a molecule of that substance. This is also the same as the **Molar Mass**, which is the mass of one mole of a substance or compound. However, Molar mass has a unit, grams per mol (g mol $^{-1}$). That simply means we are measuring the number of grams per mole of the substance.

Sometimes, you might come across **Relative Formula Mass** for a chemical substance instead of Relative Molecular Mass. Both terms mean the same thing, but Formula Mass is used for **Ionic** substance or compound. So to calculate the Relative

Formula Mass you would still add up all the atomic masses of all the ions in the Ionic compound.

What is an Ion then?

Remember we said earlier that an Atom contains electrons, neutrons and protons. Electrons have a negative charge (-ve). Protons have a Positive charge (+ve) and Neutrons have no charge. Normally, atoms would have same number of electrons and protons. So if an atom has 4 electrons you would expect it to also have 4 protons. Therefore the 4 positive (+ve) charges from the proton will cancel out the 4 negative (–ve) charges from electron. The neutron as you probably know does not contribute any charge. We therefore would usually find that atoms do not have an overall charge because of this cancelling effect.

However, this is not always the case, as sometimes due to reactions or other forms of interactions an atom might gain or lose some of either its electrons or protons. When this happens it has more of either electrons or protons. So it would now have a positive charge if it has more protons than electrons or a negative charge if it has more electrons. When it has a charge we refer to it as an **Ion**. An **Ion** is any atom or group of atoms which posses an electric charge. If this atom is part of a molecule or compound we call that compound an **Ionic** compound, because it now has a charge resulting from the atom or atoms with charge.

This brings us to **Mass Number** and **Atomic Number**. Mass number (represented by A) is simply the sum of the number of protons and neutrons in an atom, while the Atomic Number (represented by Z) is the number of protons in an atom.

We need to pay attention to these numbers as they are used to identify an element. Atoms of an element have the same atomic number or simply put, each atom of that element will contain the same number of protons.

In chemistry as you might know these two numbers are illustrated as follows;

 where X is the symbol

of the element

Based on what we have covered so far we can now say that for an Atom which has no charge, which means it has equal number of protons and electrons. We can get the number of neutrons it contains by taking away Atomic number from the Mass Number. Remember Mass number is the total of the number of protons and number of neutrons.

Sometimes we will find that atoms of the same element in their neutral state that is with no charge have the same number of protons but different number of neutrons. This phenomenon is called **Isotopy**. The different atoms are called Isotopes. We shall study this in more details when we look at Isotopy.

Now let us do some exercises on Relative Molecular Mass

EXERCISES

1. Calculate the relative molecular mass of sodium chloride, NaCl   (Atomic mass of Na=23 and Cl=35.5)

Solution

> To solve this question you will need the atomic mass of the elements. This is usually given as above or on your periodic table. Check the back of the book.

One atom of sodium chloride is made up of one atom of Na and one of Cl.

The relative molecular mass of NaCl
= 1x atomic mass of Na + 1x atomic mass of Cl
= 1x23 + 1x 35.5
= 58.5

> We multiplied by 1 because it contains one of each element.

The Relative Molecular mass of NaCl = 58.5

2. Calculate the relative molecular mass of Calcium Chloride, $CaCl_2$
(Atomic mass of Ca=40 and Cl=35.5)

Solution

One molecule of $CaCl_2$ is made up of one atom of Ca and 2 atoms of Cl.

The relative molecular mass of $CaCl_2$

= 1x atomic mass of Ca + 2x atomic mass of Cl

= 1x 23 + 2x 35.5

= 23 + 71

= 94

The relative molecular mass of $CaCl_2$ = 94

3. Calculate the relative molecular mass of calcium trioxocarbonate (iv), $CaCO_3$ (Atomic masses are Ca=40, C=12, O=16 respectively)

Solution

One molecule of $CaCO_3$ is made up of one atom of Ca, one atom of C and 3 atoms of Oxygen.

The relative molecular mass of $CaCO_3$

= 1x atomic mass of Ca + 1x atomic mass of C + 3x atomic mass of Oxygen

= 1x 40 + 1x 12 + 3x 16 = 100

The relative molecular mass of $CaCO_3$ = 100

A Mole

As mentioned earlier a mole of a substance is the amount containing as many elementary entities as the number of atoms in exactly 12g of carbon-12. This is the quantitative unit assigned to a special number of particles, to enable us have a simple presentation of reactions.

Note that:

One mole contains 6.02×10^{23} atoms

Avogadro's number (constant) = 6.02×10^{23}

The relative atomic mass of any element is equivalent to one mole of the element. Similarly, the relative molecular mass of any substance (compound or element) is equivalent to one mole of the substance.

Since one mole contains 6.02×10^{23} atoms, the relative atomic mass of any element contains 6.02×10^{23} atoms. Similarly, the relative molecular mass of any substance (compound or element) contains 6.02×10^{23} molecules of that element or compound.

Generally,

1 mole of any element or compound contains (or is equivalent to)

- 6.02×10^{23} particles
- 6.02×10^{23} atoms
- 6.02×10^{23} molecules
- 6.02×10^{23} ions

of that substance.

The **Molar mass** of any substance is the mass of one mole of that substance in grams. It is the same as the relative atomic or molecular mass.

Substance/formula	Type of particle	Atomic or molecular mass	Mass of one mole	Mass of two moles
Hydrogen atom, H	atom	1	1g	2g
Hydrogen gas, H_2	molecule	2	2g	4g
Oxygen atom, O	atom	16	16g	32g
Oxygen gas, O_2	molecule	32	32g	64g
Ozone, O_3	molecule	48	48g	96g
Sodium Chloride, NaCl	molecule	58.5	58.5g	117g
Hydrogen chloride, HCl	molecule	36.5	36.5g	73g
Sodium ion, Na^+	ion	23	23g	46g
Hydroxide ion, OH^-	ion	17	17g	34g

4. What is the mass of 3 moles of oxygen, O_2? (Atomic mass of O=16)

Solution

The mass of 1 mole of O_2 = (2x16) = 32g

Therefore, the mass of 3 moles of O_2= (3x32) = 96g

The mass of 3 moles of O_2 = 96g

5. How many atoms are there in 8g of oxygen?

(1 mole = 6.02×10^{23}, O = 16)

Solution

The mass of 1 mole of oxygen is 16g, and
Since 1 mole of any element contains 6.02×10^{23} atoms,
16g of oxygen contains 6.02×10^{23} atoms
Therefore, 8g of oxygen will contain n atoms
Cross-multiply and solve for n

$$n = \frac{8 \times 6.02 \times 10^{23}}{16}$$

$n = 3 \times 10^{23}$ atoms
8g of oxygen contain 3×10^{23} atoms.

6. How many moles are there in 30g of calcium trioxocarbonate(iv), $CaCO_3$ (By now you may have known that these are the Atomic masses of the elements Ca= 40, C= 12, O= 16)

Solution

First, we calculate the molar mass (i.e. the mass of 1 mole) of $CaCO_3$
One mole of $CaCO_3$ is made up of 1 atom of Ca, 1 atom of C and 3 atoms of O.
Molar mass of $CaCO_3$ = $1 \times 40 + 1 \times 12 + 3 \times 16$
= 100g

Since, 100g of $CaCO_3$ contains 1 mole
30g of $CaCO_3$ will contain n mole
Cross multiply and solve for n

$n = \dfrac{30 \times 1}{100} = 0.3$

Therefore, 30g 0f $CaCO_3$ contains 0.3 mole.

7. Calculate the percentage by mass of hydrogen in water. (H= 1, O= 16)

Solution

Remember, when you add up the mass of the atoms in a compound it gives you the mass of one mole of that compound or substance, which is the molar mass. So to work out the percentage by mass of an element in a compound we simply divide the mass of the atoms of that element by the mass of one mole of that compound and multiply by hundred. It is just simple percentage calculation, like saying there 10g of sand and 90g of stones in a bucket what is the percentage of sand in the bucket.

Percentage by mass of an element in a compound = (total mass of element in the compound/molar mass of compound) x 100

The molecular formula of water is H_2O.
One mole of H_2O is made up of 2 moles of hydrogen, and 1 mole of oxygen.

The molar mass of H_2O = 2 x 1 + 1 x 16

= 18g

18g of H_2O contain 2g of hydrogen.

Therefore, the percentage by mass of hydrogen in H_2O

= (2/18) x 100

= 11.1%

The percentage by mass of hydrogen in H_2O = 11.1%

8. Calculate the percentage by mass of sulphur in tetraoxosulphate(vi) acid, H_2SO_4. (H= 1, S = 32, O= 16)

Solution

The molecular formula of tetraoxosulphate (VI) acid is H_2SO_4. A mole of H_2SO_4 is made up of 2 moles of H, 1 mole of S and 4 moles of O.

The molar mass of H_2SO_4 = 2 x1 + 1 x 32 + 4 x 16

= 98g

Therefore, 98g of H_2SO_4 contain 32g of sulphur.

The percentage by mass of sulphur in H_2SO_4 = (32/ 98) x 100

= 32.7%

The percentage by mass of sulphur in H_2SO_4 = 32.7%

MOLECULAR AND EMPIRICAL FORMULA

The molecular formula of a compound gives the exact number of moles of atoms of the component elements in a mole of the compound. While, the empirical formula of a compound gives the (simple) ratio of the component elements. To better illustrate this let us look at glucose

The Molecular formula for glucose is $C_6H_{12}O_6$

This formula represents a molecule of glucose.

However, its empirical formula is CH_2O. This is the simple ratio of the elements which make up a molecule of glucose. What this means is that for every atom of Carbon in glucose we have two atoms of Hydrogen and also one atom of Oxygen.

If you have a substance with a molecular formula of X_5Y_{20}, you may have guessed the empirical formula would be XY_4

Therefore we could easily say in simple mathematical terms that

Molecular formula = (Empirical Formula) x n

Molecular formula = (XY_4) n

n in the above case is 5.

Molecular formula = $(XY_4)5 = X_5Y_{20}$

Usually, we calculate the mole ratio of an element by dividing the reacting mass by its Molar mass.

EXERCISES

1. A substance has the molecular formula C_4H_6. What is its empirical formula?

Solution

In each molecule of the substance there are 4 atoms of Carbon for every 6 atoms of Hydrogen. That means we could further say we have 2 carbon atoms for every 3 Hydrogen atoms.

> Since, if we go any further we would end up with a fraction, which is for every 1 atom of Carbon we have 1.5 of Hydrogen, we don't go any further to simplify. You always want the ratio to be whole numbers

Therefore the Empirical Formula for this substance is C_2H_3

2. A substance is found to have a relative molecular mass of 114. If its empirical formula is C_4H_9, what is its Molecular formula?

Solution

We have been given the empirical formula as well as the molecular mass. Remember, earlier we said the empirical formula is the simple ratio of the elements. To get the molecular formula we need to find n,

Molecular Formula = (Empirical Formula, C_4H_9) x n

Now we can find the Molecular formula by writing a mass equation for the molecule.

The mass of a molecule of the substance is 114, and we are trying to find the formula for this molecule. (Remember Atomic mass of Carbon = 12 and Hydrogen is 1)

$114 = (C_4H_9) n$

$114 = ((12 \times 4) + (1 \times 9))n$

$114 = (48 + 9) n$

$114 = 57n$ (to get n, we divide both sides by 57)

$114/57 = n$

$2 = n$

Since we now have n, we can easily work out the molecular formula,

Molecular Formula = $(C_4H_9)2$

Molecular Formula = C_8H_{18}

3. Calculate the molecular formula for a compound which on analysis yielded 4.00g of carbon, 0.68g of hydrogen and 5.34g of oxygen. If the relative molecular mass of the compound is 60. (C = 12, H = 1, O = 16)

Solution

To solve a question like this one, we first find the empirical formula by the following process, employing a table. If you think about it for every 4.00g of Carbon, there was 0.68g of Hydrogen and 5.43g of Oxygen. You could relate this to the explanation we gave for empirical formula earlier.

Element	Carbon,C	Hydrogen,H	Oxygen,O
Reacting mass (*mass yielded*)	4.00g	0.68g	5.34g
Mole ratio of element = reacting mass/molar mass	4.00/12 = 0.33	0.68/1 = 0.68	5.34/16 = 0.33
Mole ratio of element/smallest mole ratio	0.33/0.33 = 1	0.68/0.33 = 2	0.33/0.33 = 1
Ratio	1	2	1

N.B. Always round the ratio to the nearest whole number.

Therefore,

The empirical formula of the compound is $C_1H_2O_1 = CH_2O$

Remember Molecular formula = (Empirical Formula) x n

Now let the molecular formula of the compound be $(CH_2O)n$

We are already given the relative molecular mass as 60 so we could easily say that the mass of

$(CH_2O)n = 60$

We now have a mass equation which we could easily solve by substituting the mass of all the elements.

(1 x atomic mass of C + 2 x atomic mass of H + 1 x atomic mass of O) = 60

$(12 + 2 + 16)n = 60$

30n = 60

n = 60/30 = 2

Substitute n in the empirical formula, $(CH_2O)n$

$(CH_2O)2 = C_2H_4O_2$

The molecular formula of the compound is $C_2H_4O_2$

4. A compound contains 40.0% carbon, 6.7% hydrogen and 53.3% oxygen. If the molar mass of the compound is 180, find the molecular formula?

Solution

In this case to get the mole ratio of elements we divide the percentage by the respective molar mass.

	C	H	O
Percentage	40%	6.7%.	53.3%
Mole ratio of element= percentage/molar mass	40.0/12 =3.33	6.7/1 = 6.7	3.3/16 = 3.33
Mole ratio of element/ smallest mole ratio	3.33/3.33 = 1	6.7/3.33 = 2	3.33/3.33 = 1
Ratio	1	2	1

The empirical formula of the compound = CH_2O

Let the molecular formula of the compound = $(CH_2O)n$

$(CH_2O)n = 180$

$(1 \times 12 + 2 \times 1 + 1 \times 16)n = 180$

$30n = 180$

$n = 180/30 = 6$

Substituting the value of n in $(CH_2O)n$

$(CH_2O)6 = C_6H_{12}O_6$

Therefore, the molecular formula of the compound is $C_6H_{12}O_6$

Do you recollect what this compound is called?

5. What is the percentage by mass of oxygen in $Al_2(SO_4)_3.2 H_2O$?

(Al = 27, S = 32, O = 16)

Solution

First, we calculate the molar mass of the compound.

$= 2 \times 27 + 3(32 + 4 \times 16) + 2(2 \times 1 + 16)$

$= 54 + 288 + 36$

$= 378g$

The total mass of oxygen in the compound $= 3(4 \times 16) + 2 \times 16 = 224$

So, the percentage by mass of oxygen $= (224/378) \times 100$

$= 59.26\%$

The percentage by mass of oxygen in $Al_2(SO_4)_3.2 H_2O$ is 59.26%

6. A compound contains 31.91% potassium, 28.93% chlorine and the rest oxygen. What is the chemical formula of the compound?
(K = 39, Cl= 35.5, O= 16)

Solution

Chemical formula here is simply the same thing as molecular formula.

First, let us determine the percentage by mass of oxygen in the compound

The percentage by mass of oxygen in the compound = 100% - (%potassium + %chlorine)

= 100% - (31.91% + 28.93%) = 39.16%

Percentage by mass of oxygen = 39.16%

	K	Cl	O
Percentage	31.91%	28.93%	39.16%
Mole ratio of element	31.91/39 = 0.82	28.93/35.5 = 0.82	39.16/16 = 2.5
Mole ratio of element/smallest mole ratio	0.82/0.82 = 1	0.82/0.82 = 1	2.5/0.82 =3
Ratio	1	1	3

The chemical formula of the compound = $KClO_3$

ISOTOPY

This is the phenomenon where we have atoms of the same element, having the same atomic number but different mass numbers.

Remember, Atomic number is the number of protons in the nucleus of an element and Mass number is the sum of the number of protons and neutrons in the nucleus of an atom.

To determine the relative molecular mass of a sample of an element, which contains a mixture of the isotopes.

The relative atomic mass = [(percentage abundance of isotope A/100) x Weight or mass number of A] + [(percentage abundance of isotope B/100) x Weight or mass number of B]

The relative atomic mass of an element, which exhibits isotopy, is the average mass of its various isotopes in a sample of the element.

EXERCISES

1. Chlorine, consisting of two isotopes of mass numbers 35 and 37, has an atomic mass of 35.5. The relative abundance of the isotope of mass number 37 is what?

Solution

We know relative atomic mass = [(percentage abundance of isotope ^{35}Cl/100) x mass number of ^{35}Cl] + [(percentage abundance of isotope ^{37}Cl/100) x mass number ^{37}Cl]

In this case we have two unknowns, but if you consider it, the percentage abundance of ^{35}Cl + percentage abundance of ^{37}Cl = 100% or 1 if we express as decimals.

Let us assume the percentage abundance of ^{37}Cl = n, so percentage abundance of ^{35}Cl will be 1 – n.

Substituting, values we have an equation.

35.5 = [(1-n) x 35] + [(n) x 37]

If we open up the brackets

35.5 = 35 – 35n + 37n

Rearranging this equation we have

35.5 – 35 = 37n – 35n

0.5 = 2n

n = 0.5/2

n = 0.25

If we revert back to hundred this will become 0.25 x 100 = 25%

Therefore, percentage abundance of ^{37}Cl is 25%.

2. Calculate the average atomic weight for Nitrogen, made up 99.63% of ^{14}N with atomic mass of 14.003074 and 0.37% of ^{15}N with atomic mass of 15.000108

Solution

The relative atomic mass = [(percentage abundance of isotope A/100) x Weight or mass number of A] + [(percentage abundance of isotope B/100) x Weight or mass number of B]

The relative atomic mass is the same as the average atomic weight.

So we could find the average atomic weight by substituting in the above equation all the values we have been given.

Average atomic weight = [(99.63/100) 14.003074] + [(0.37/100) 15.000108]

= 13.978827 + 0.0555003996

= 14.03432

Average atomic weight of this sample of Nitrogen = 14.03432

PRACTICE QUESTIONS

Now can you practice what we have done so far..

1. 2.50g of a compound containing only phosphorus and chlorine was analyzed and found to contain 1.94g of chlorine. What is the empirical formula of the compound?

2. A compound containing carbon, hydrogen and nitrogen was found to contain 53.1% carbon, 15.9% hydrogen. Determine the empirical formula of the compound.

3. 5.0g of an organic compound on analysis showed that it composed of 27.2g carbon and 0.48g hydrogen. Use the data to determine the empirical formula of the compound. If the molecular weight of the compound is 88. What is the molecular formula of the compound?

MASS AND VOLUME RELATIONSHIPS IN REACTIONS

Solving mathematical problems in chemistry requires a sound knowledge of writing proper chemical formulae of substances and balanced chemical equations for reactions. We shall not be covering balancing chemical reactions in this book.

From a balanced chemical equation, one can deduce the stoichiometry of the reaction. This is the key to solving problems. The stoichiometry of a reaction is the mole ratio in which reactants combine and products are formed.

In a balanced chemical equation, the numerical coefficients represent the number of moles of reactants and products. From these coefficients, we get the mole ratio of the reactants and products in a reaction.

Let us consider the following equations

1. $HCl(aq) \rightarrow H^+(aq) + Cl^-(aq)$

In this ionization equation, one mole of hydrochloric acid produces one mole of hydrogen ion and one mole of chloride ion in solution.

2. $H_2SO_4(aq) \rightarrow 2\ H^+(aq) + SO_4^{2-}(aq)$

Here, one mole of tetraoxosulphate(vi) acid produces two moles of hydrogen ions and one mole of tetraoxosulphate(vi) ion in solution.

3. $NaOH(aq) + HCl(aq) \rightarrow NaCl(aq) + H_2O(l)$

	NaOH	HCl	NaCl	H₂O
No of Moles	1	1	1	1
Mole Ratio	1 :	1 :	1 :	1
Molar mass	40g	36.5g	58.5g	18g
Reacting mass	40g	36.5g	58.5g	18g

4. $Zn(s) + 2HCl(aq) \rightarrow ZnCl_2(aq) + H_2(g)$

	Zn	HCl	ZnCl₂	H₂
No of moles	1	2	1	1
Mole ratio	1 :	2 :	1 :	1
Molar mass	65g	36.5g	100.5g	2g
Reacting mass	65g	(2 x 36.5)g	100.5g	2g

Reacting mass = Number of mole x molar mass

In equation 3, one mole of NaOH reacts with one mole of HCl to produce one mole of NaCl and one mole of water.

In equation 4, one mole of Zn reacts with two moles of HCl to produce one mole of zinc chloride and one mole of hydrogen.

Mole ratios and Mass relationships

In equation 3,

40g of NaOH reacts with 36.5g of HCl, to produce 58.5g of NaCl and 18g of water.

In equation 4, 65g of Zn reacts with 73g of HCl to produce 100.5g of zinc chloride and 2g of hydrogen gas.

We can calculate the reacting mass or the number of moles of any one of the substances in a reaction, from the balanced equation of the reaction, if the reacting mass or the number of mole of the others is given.

You must carryout analyses of a question before attempting to solve it. Look at what you are given and what you require to get your answer. Then think of how to get those values required to get the answer.

EXERCISES

1. Calculate the number of moles of water, which could be obtained from 30g of limestone, $CaCO_3$, in the presence of excess hydrogen chloride, HCl

(Ca=40, C=12, O=16, H = 1, Cl = 35.5)

Solution

First, we write a balanced equation for the reaction.

$CaCO_3(s) + 2HCl(aq) \rightarrow CaCl_2(aq) + H_2O(l) + CO_2(g)$

After writing the equation, you ask yourself which of the compounds you are concerned with, then the form of each of the compounds you are dealing with, is it mole, mass or volume.

In this question, we are concerned with the mass of $CaCO_3(s)$ and the number of mole of water. From the equation,

One mole of $CaCO_3(s)$ yields one mole of water.

That is, 40g of $CaCO_3(s)$ yields one mole of water

Therefore, 30g of $CaCO_3(s)$ will yield y mole of water

Cross-multiply and solve for y

Y = 30 x 1/40   = 0.75mole

30g of limestone will yield 0.75mole of water.

2. What mass of carbon (iv) oxide will be produced on burning 100g of ethyne, C_2H_2.
(C=12, O=16, H = 1)

Solution

Equation for the reaction,

$2 C_2H_2(g) + 5O_2 \rightarrow 4CO_2(g) + 2 H_2O(g)$

Here we are concerned with the mass of ethyne and the mass of CO_2.

From the equation

2 moles of ethyne yields 4 moles of CO_2 gas on burning.

That is, (2 x 26)g of ethyne yields (4 x 44)g of CO_2 on burning.

Therefore, 100g of ethyne will yield yg of $CO_2(g)$ on burning.

We cross multiply and solve for y

Y = 100 x 4 x 44
       ──────────
        2 x 26   = 338.5g

100g of ethyne will yield 338.5g of $CO_2(g)$ on burning.

VOLUME RELATIONSHIPS FOR GASES

From Avogadro's law, the molar volume of any gas is the volume occupied by one mole of that gas at S.T.P. and is numerically equal to $22.4dm^3$. That is one mole of all gases occupy a volume of $22.4dm^3$ at s.t.p.

Equal volumes of gases contain the same number of molecules at a given temperature and pressure.

We can calculate the volume of gases involved in a chemical reaction from the balanced equation. The coefficients of gaseous reactants and products give the mole relations as well as the volume relations of the gases. Provided the gases are under the same conditions (pressure and temperature). For instance,

$N_2(g) + 3H_2(g) \rightarrow 2NH_3(g)$

In the above reaction, one mole of nitrogen reacts with three moles of hydrogen to produce two moles of ammonia. It follows that one molar volume of nitrogen reacts with three molar volumes of hydrogen to produce two molar volumes of ammonia.

That is $22.4dm^3$ of nitrogen reacts with $3 \times 22.4dm^3$ of hydrogen to produce
$2 \times 22.4dm^3$ of ammonia.

EXERCISES

1. Find the volume of oxygen produced by one mole of potassium trioxochlorate (v) at s.t.p. in the following reaction.
$2KClO_3(s) \rightarrow 2KCl(s) + 3O_2(g)$

Solution

In the question, we are concerned with the number of mole of $KClO_3$ and the volume of oxygen.

From the equation,

2 moles of $KClO_3(s)$ yields 3 moles of $O_2(g)$

That is, 2 moles of $KClO_3(s)$ yields 3 x 22.4dm³ of oxygen.

Therefore, 1 mole of $KClO_3(s)$ will yield ydm³ of oxygen.

Solve for y

$Y = \underline{1 \times 3 \times 22.4}$
    $\quad\quad 2 \quad\quad = 33.6\ 4dm^3$

1 mole of $KClO_3(s)$ will yield 33.6 4dm³ of oxygen.

2. Find the mass of sodium trioxocarbonate(iv) needed to give 22.4dm³ of carbon(iv)oxide at s.t.p. in this reaction.
$Na_2CO_3(s) + 2HCl(aq) \rightarrow 2NaCl(aq) + CO_2(g) + H_2O(l)$
[Na = 23, O = 16, C = 12, GMV = 22.4dm³]

Solution

In the question we are concerned with the mass of $Na_2CO_3(s)$ and the volume of $CO_2(g)$. From the equation of the reaction One mole of $Na_2CO_3(s)$ yields one mole of $CO_2(g)$ on reacting

(molar mass of Na_2CO_3 = 106)

Therefore, **106g of Na_2CO_3 (s) yields 22.4dm^3 of CO_2(g)**

3. Calculate the mass of $ZnSO_4$ produced when excess of $ZnCO_3$ is added to 50.0cm^3 of 4.00moldm^{-3} H_2SO_4.
The equation for the reaction is,

$H_2SO_4(aq) + ZnCO_3(s) \rightarrow ZnSO_4(aq) + CO_2(g) + H_2O(l)$

[$ZnSO_4$ = 161gmol^{-1}]

Solution

First let us find the mass of H_2SO_4 in 50 cm^3 of 4.00moldm^{-3} H_2SO_4

We have to convert the concentration to g dm^{-3}.

g dm^{-3} = moldm^{-3} x molar mass

= 4.00 x 98 = 392 gdm^{-3}

1 dm^3 (1000cm^3) of solution contains 392g of H_2SO_4

Therefore, 50 cm^3 of solution will contain ng of H_2SO_4

Solving for n,

n = $\underline{50 \times 392}$
      1000   ≈ 19.6g

The reacting mass of H_2SO_4 that reacted was 19.6g

From the equation of reaction,

1 mole of H_2SO_4 yields 1 mole of $ZnSO_4$

That is, 98g of H_2SO_4 yields 161g of $ZnSO_4$

Therefore, 19.6g of H_2SO_4 will yield ng of $ZnSO_4$

solve for n,

n = $\frac{19.6 \times 161}{98}$ = 32.2g of ZnSO$_4$

32.2g of ZnSO$_4$ is produced

4. A chip used in a microcomputer contains 5.72×10^{-3} g of silicon. Calculate the number of silicon atoms in the chip. [Si = 28, Avogadro's constant = 6.02×10^{23}]

Solution

Remember, 1 mole of an element contains 6.02×10^{23} atoms
Hence, 1 mole of silicon contains 6.02×10^{23} atoms
That is, 28g of silicon contains 6.02×10^{23} atoms
Therefore, 5.72×10^{-3} g will contain m no of atoms
Solve for m,

m = $\frac{5.72 \times 10^{-3} \times 6.02 \times 10^{23}}{28}$ = 1.23×10^{20} atoms

There are 1.23×10^{20} atoms of silicon in the chip.

5. Bleaching powder reacts with dilute hydrochloric acid according to the following equation, CaOCl$_2$ (s)+ 2HCl(aq) → CaCl$_2$(aq) + Cl$_2$(g).
Calculate the mass of bleaching powder that will produce 400 cm^3 of chlorine at 25°C and a pressure of 1.20×10^5 NM^{-2}

[O = 16, Cl = 35.5, Ca = 40, GMV= 22.4dm^3 at s.t.p., standard pressure=1.01 x 10^5 NM^{-2}]

Solution

First, we convert the volume of chlorine produced at 25°C and 1.20 x 10^5 NM^{-2} to the volume at standard temperature and pressure.

Using the general gas equation,

$$\frac{P_1V_1}{T_1} = \frac{P_2V_2}{T_2}$$

> Initial Pressure, P_1
> Final Pressure, P_2
> Initial Volume, V_1
> Final Volume, V_2
> Final Temperature, T_2
> Initial Temperature, T_1

Initial Pressure, $P_1 = P_2 = 1.20$ x 10^5 NM^{-2}
Initial Volume, $V_1 = ?$
Final Volume, $V_2 = 400$ cm^3
Final Temperature, $T_2 = 273 + 25 = 298$K
Initial Temperature, $T_1 = 273$K(standard temperature)

$$V_1 = \frac{P_2V_2T_1}{P_1T_2}$$

$$= \frac{1.20 \times 10^5 \text{ NM}^{-2} \times 400 \times 273}{1.01 \times 10^5 \text{ NM}^{-2} \times 298}$$

$$= 435.37 \text{ cm}^3$$

The volume of Cl$_2$ produced at standard temperature and pressure is 435.37 cm^3, if we convert this to dm^3, it becomes 435.37/1000 = 0.43537 dm^3

To calculate the mass of bleaching powder required, we look at the mole ratio of $CaOCl_2$ which is reacting and the Cl_2 produced

From the equation 1 Mole of $CaOCl_2$ yields 1 Mole of Cl_2
This means that
The molar mass of $CaOCl_2$ (40 + 16 +2x35.50) yields the molar volume of Cl_2 (22.4 dm^3)
127 g of $CaOCl_2$ yields 22.4 dm^3 of Cl_2,
Therefore n g of $CaOCl_2$ will yield 0.43537 dm^3 of Cl_2
If we cross multiply and solve for n
22.4n = 127 x 0.43537
n = (127 x 0.43537)/22.4 = 55.29199/22.4
n = 2.468 g
2.468 g of bleaching powder will yield 435.37 cm^3 of Cl_2 at standard temperature and pressure, and 400 cm^3 of chlorine at 25°C and a pressure of 1.20 x 10^5 NM^{-2}

6. What amount of copper is deposited when 13.0g of zinc reacts with excess copper (ii) tetraoxosulphate (vi) solution according to the following equation?
$Zn(s) + CuSO_4 (aq) \rightarrow ZnSO_4 (aq) + Cu(s)$
[Cu=63.5, Zn=65]

Solution
From the equation,

1 mole of zinc deposits 1 mole of Cu
that is, 65g of zinc deposits 63.5g of Cu
Therefore, 13.0g of zinc will deposit ng of Cu
solve for n,

n = $\frac{13 \times 63.5}{65}$ = 12.7g

13.0g of zinc deposits 12.7g of Cu

7. What volume of oxygen at standard temperature and pressure, S.T.P.. would react with carbon to form 4.40g of CO_2 according to the following equation?

$C(s) + O_2(g) \rightarrow CO_2(g)$

[O=16m C=12, GMV at s.t.p = $22.4dm^3$]

Solution

We are concerned with the volume of oxygen and the mass of CO_2 (g).

From the equation of the reaction, 1 mole of oxygen yields 1 mole of CO_2 (g) .(molar mass of $CO_2(g)$ = 44g)

Hence, $22.4dm^3$ of oxygen yields 44g of $CO_2(g)$

Therefore, m volume of oxygen will yield 4.40g of $CO_2(g)$

Cross multiply and solve for m,

m = $\frac{22.4 \times 4.4}{44}$ = 2.24 dm^3

2.24 dm^3 of oxygen will form 4.4g of $CO_2(g)$

8. What is the percentage by mass of sulphur in $Al_2(SO_4)_3$? [$Al_2(SO_4)_3 = 342 \text{gmol}^{-1}$, $S = 32$]

Solution

The percentage by mass of sulphur will be = [mass of sulphur/molar mass of compound] x 100
= [(3x 32)/342] x 100 = 28.1%
The percentage by mass of sulphur in $Al_2(SO_4)_3$ is 28.1%

9. Xg of a pure sample of iron (ii) sulphide reacted completely with excess dilute hydrochloric acid to give 3.20g of iron(ii)chloride according to the following equation, $FeS(s) + 2HCl(aq) \rightarrow FeCl_2(aq) + H_2S(s)$

Solution

In the question, we are concerned with the mass of FeS and the mass of $FeCl_2$. From the equation of the reaction,
1 mole of FeS produces 1 mole of $FeCl_2$
that is, 91.5g of FeS produces 127g of $FeCl_2$
Therefore, Xg of FeS will produce 3.2g of $FeCl_2$
Solve for n,
X = 91.5 x 3.2
      127
X = 2.31g

10. When powdered magnesium is heated to redness in a stream of nitrogen, magnesium nitride (Mg_3N_2) is formed.

(i) Write an equation for the reaction.

(ii) Hence, calculate the amount in mole of magnesium nitride that can be obtained from 3.0g of magnesium. [Mg = 24]

$3 Mg(s) + N_2(g) \rightarrow Mg_3N_2(s)$

We are concerned with the mole of Mg_3N_2 and the mass of Mg.

From the equation of the reaction,

3 moles of Mg produces 1 mole of Mg_3N_2

Hence, (3 x 24) g of Mg produces 1 mole of Mg_3N_2
Therefore, 3g of Mg will produce n mole of Mg_3N_2
Solve for n,

$n = \dfrac{3 \times 1}{3 \times 24} = 0.042$ mole

Therefore, 0.042 mole of Mg_3N_2 can be obtained from 3g of magnesium.

11. One mole of a compound $M(HCO_3)_2$ has a mass of 162g. Calculate the relative atomic mass of M. [H=1, C=12, O=16]

Solution

The mass of one mole of $M(HCO_3)_2$ is its molar mass, that is, 162g. Hence, we can determine the relative atomic mass of M thus,

Molar mass of $M(HCO_3)_2$ = atomic mass of M + (atomic mass of H + atomic mass C + 3x atomic mass of O) x 2

$162 = M + (1 + 12 + 48)2$

$162 = M + 2 \times 61$

$162 = M + 122$

$M = 162 - 122$

$M = 40$

The relative atomic mass of M = 40

12. What volume of hydrogen is produced at s.t.p. when 2.60g of zinc reacts with excess HCl according to the following equation?

$Zn(s) + 2HCl(aq) \rightarrow ZnCl_2(aq) + H_2 (g)$

[Zn=65, GMV of gas = $22.4dm^3$]

Solution

From the equation, 1 mole of zinc yields 1 mole of hydrogen.
Hence, 65g of Zn yields $22.4dm^3$ of hydrogen
Therefore, 2.6g of Zn will yield n dm^3 of hydrogen
Solve for n,
$n = 2.6 \times 22.4/65 = 0.896 dm^3$
2.6g of Zn yields $0.896 dm^3$ of hydrogen

13. $30 cm^3$ of hydrogen at s.t.p. combines with $20 cm^3$ of oxygen to form steam according to the following equation,

$2 H_2 (g) + O_2(g) \rightarrow 2H_2O (g)$

Calculate the total volume of the gaseous mixture at the end of the reaction.

Solution

The volume of the gaseous mixture will be the volume of water formed plus volume of any excess of the reactants left. So we find the reactant, which was in excess first.

From the equation,

2 moles of hydrogen reacts with 1 mole of oxygen to form 2 moles of water.

Since, 2 x 22400 cm^3 of hydrogen combines with 22400 cm^3 of oxygen,

30 cm^3 of hydrogen will combine with n cm^3 of oxygen,

Solve for n,

n = $\dfrac{30 \times 22400}{2 \times 22400}$ = 15 cm^3

So 30 cm^3 of hydrogen only combines with 15 cm^3 of oxygen, leaving an excess of oxygen, which is 20 – 15 = 5 cm^3

Since 2 moles of hydrogen forms 2 moles of water,

30 cm^3 of hydrogen will also form 30 cm^3 of water.

Therefore the total volume of the gaseous mixture at the end of the reaction

= volume of water formed + volume of excess oxygen

= 30 + 5 = 35 cm^3

Alternative Solution

Note that the total volume of the gaseous mixture would be the volume of the product formed and the volume of the excess gas. Applying Gay-Lussac's law of combining volumes. From the equation, 2 moles of hydrogen requires 1 mole of oxygen to form 2 moles of water.

Therefore, 30 cm^3 of hydrogen will react with only 15 cm^3 of oxygen to form 30 cm^3 of water.

Hence the volume of excess oxygen = 20 − 15 = 5 cm^3

So, if we add this to the volume of the product, water formed, we have

Total volume of gaseous mixture = 30 + 5 = 35 cm^3

14. If the mass of one molecule of the gas G is 2.19×10^{-22}g, determine the molar mass of G [Avogadrro's constant = 6.02×10^{23}]

Solution

1 mole of a substance contains 6.02×10^{23} molecules

Since, the mass of 1 molecule of G is 2.19×10^{-22}g,

then, the mass of 6.02×10^{23} molecules will be ng

solve for n,

$$n = \frac{6.02 \times 10^{23} \times 2.19 \times 10^{-22}}{1} = 132g$$

The molar mass of M = 132g

15. Calculate the number of mole of electrons involved in the oxidation of 2.8g of iron fillings to iron(ii) ions. [Fe = 56]

Solution

The oxidation equation is Fe(s) → Fe^{2+}(aq)

From the equation, 2 moles of electron are involved in the oxidation of 1 mole of Fe to Fe^{2+}(aq).

Hence, 2 moles of electron are involved in the oxidation of 56g of Fe,

Therefore, n mole will be involved in the oxidation of 2.8g of Fe

Solve for n,

$n = \dfrac{2 \times 2.8}{56} = 0.1$

0.1 mole of electron will be involved in the oxidation of 2.8g of Fe fillings to Fe (ii) ions.

16. 20g of copper(ii)oxide was warmed with 0.05 mole of tetraoxosulphate(vi) acid. Calculate the mass of copper(ii)oxide that was in excess. The equation for the reaction;

$CuO(s) + H_2SO_4(aq) \rightarrow CuSO_4(aq) + H_2O(l)$

[O = 16, Cu = 64]

Solution

From the equation, usually, 1 mole of CuO reacts with 1 mole of H_2SO_4(aq). That is 80g of CuO reacts with 1 mole of H_2SO_4(aq).

Now to find the mass of CuO that was in excess, we have to first find the mass of CuO which reacted with 0.05 mole of H_2SO_4(aq).

1 mole of $H_2SO_4(aq)$ reacts with 80g of CuO
Therefore, 0.05 mole of $H_2SO_4(aq)$ will react with ng of CuO
Solve for n,
n = 0.05 x 80/1 = 4g
Since, 0.05 mole of $H_2SO_4(aq)$ reacts with only 4g of CuO.
The mass of excess CuO = 20g – 4g = 16g

17. Calculate the volume of oxygen that was in excess if 150 cm^3 of carbon (ii) oxide was burnt in 80 cm^3 of oxygen according to the following equation; $2CO(g) + O_2(g) \rightarrow 2CO_2(g)$

Solution

From the equation, usually, 2 moles of CO reacts with 1 mole of oxygen.
That is, 2 x 22.4dm^3 of CO reacts with 22.4dm^3 of oxygen.
To find the volume of oxygen that was in excess, we have to first, find the volume of oxygen that would react with 150cm^3 of CO.
2 x 22.4dm^3 of CO reacts with 22.4dm^3 of oxygen.
Therefore, 0.15dm^3 of CO will react with n dm^3 of oxygen.
Solve for n,

n = $\underline{0.15 \times 22.4}$
          2 x 22.4   = 0.075 dm^3 = 75cm^3

Since, 150 cm^3 of CO reacts with only 75 cm^3 of oxygen. The volume of oxygen that was in excess = 80 – 75 = 5 cm^3

18. Calculate the volume of chlorine at s.t.p., that would be required to react completely with 3.70g of dry slaked lime according to the following equation

$Ca(OH)_2(s) + Cl_2(g) \rightarrow CaOCl_2(s) + H_2O(l)$

[H = 1, O = 16, Ca = 40, molar volume of gases = 22.4dm^3

Solution

In this question we are only concerned with the volume of chlorine and the mass of slaked lime. From the equation, 1 mole of slaked lime reacts completely with 1 mole of chlorine.
[molar mass of slaked lime, $Ca(OH)_2$ = 74g]
Hence, 74g of $Ca(OH)_2$ reacts completely with 22.4dm^3 of chlorine.
Therefore, 3.7g of $Ca(OH)_2$ will react completely with n dm^3 of chlorine
Solve for n,
n = 3.7 x 22.4/74 = 1.12dm^3
3.7g of slaked lime reacts completely with 1.12dm^3 of chlorine.

19. A metal, M with relative atomic mass 56 forms an oxide with formula X_2O_3. How many grams of the metal will combine with 10g of oxygen?

Solution
First, we need to write a balanced equation for the reaction.

$4X(s) + 3O_2(g) \rightarrow 2X_2O_3(s)$

From the equation, 4 moles of the metal reacts with 3 moles of oxygen,

hence, (4 x 56)g of the metal combines with (16 x 6)g of oxygen.

Therefore, ng of the metal will combine with 10g of oxygen.

Solve for n,

$$n = \frac{4 \times 56 \times 10}{6 \times 16} = 23.3g$$

23.3g of the metal M will combine with 10g of oxygen

20. Ethane burns completely in oxygen according to the following equation, $C_2H_6(g) + 7/2 O_2(g) \rightarrow 2CO_2(g) + 3H_2O(l)$
How many moles of carbon (iv) oxide will be produced when 6.0g of ethane are completely burnt in oxygen?

Solution

We are only concerned with the number of moles of carbon (iv) oxide and the mass of ethane. From the equation, one mole of ethane produces 2 moles of $CO_2(g)$ [molar mass of ethane = 30g]

So, 30g of ethane produces 2 moles of $CO_2(g)$

Therefore, 6.0g of ethane will produce n mole of $CO_2(g)$

Solve for n,

n = 6.0 x 2/30 = 0.4 mole

0.4 mole of CO_2 (g) is produced when 6.0g of ethane burns completely.

21. 39.8g of a mixture of potassium chloride, KCl and potassium trioxochlorate (v), $KClO_3$, were heated to a constant mass. If the residue weighed 28.9g. What was the percentage mass of the potassium chloride in the mixture? [K = 39, Cl = 35.5, O = 16]

Solution
During the heating, KCl is not decomposed. But $KClO_3$, decomposes on heating. So, it is $KClO_3$, that is responsible for any loss in mass. Which would be as a result of the oxygen escaping as it is decomposes.
The loss in mass = 39.8 – 28.9 = 10.9g
So the mass of oxygen produced by $KClO_3$, is 10.9g. Now let us find the mass of $KClO_3$, which produced that amount of oxygen, and that will give us the mass of $KClO_3$ in the mixture.
The equation for the decomposition of $KClO_3$,
$2 KClO_3 \rightarrow 2KCl(s) + 3O_2 (g)$
From the equation, 2 moles of $KClO_3$ yields 3 moles of oxygen on decomposition [molar mass of $KClO_3$ = 122.5g]
So, (2 x 122.5)g of $KClO_3$ yields (3 x 32)g of oxygen
therefore, ng of $KClO_3$ will yield 10.9g of oxygen

Solve for n,

n = $\dfrac{10.9 \times 2 \times 122.5}{3 \times 32}$ = 27.8g

So the mass of $KClO_3$ in the mixture was 27.8g.

Therefore, the mass of KCl in the mixture was 39.8 – 27.8 = 12g

The percentage by mass of KCl in the mixture

= [mass of KCl/mass of mixture] x 100

= (12/39.8) x 100 = 30.2%

22. 12.5g of zinc trioxocarbonate (iv), $ZnCO_3$, was heated very strongly to a constant mass and the residue treated with excess hydrochloric acid, HCl. Calculate the mass of zinc chloride, $ZnCl_2$, that would be obtained.

[Zn = 65]

Solution

When $ZnCO_3$ is heated it decomposes to form zinc oxide and carbon(iv)oxide according to the following equation,

$ZnCO_3$ (s) → ZnO(s) + CO_2 (g)

From the equation, 1 mole of $ZnCO_3$ yields 1 mole of ZnO

[Molar mass of $ZnCO_3$ = 125, ZnO = 81]

That is, 125g of $ZnCO_3$ yields 81g of ZnO

Therefore, 12.5g of $ZnCO_3$ will yield ng of ZnO

Solve for n,

n = 12.5 x 81/125 = 8.1g

The residue that reacted with the acid was 9.1g of ZnO. The equation for the reaction is,

ZnO(s) + 2HCl(aq) → ZnCl$_2$ (aq) + H$_2$O(l)

Here we are concerned with the mass of ZnO and the mass of ZnCl$_2$

From the equation 1 mole of ZnO forms 1 mole of ZnCl$_2$

[Molar mass of ZnCl$_2$ = 136]

So, 81g of ZnO yields 136g of ZnCl$_2$

Therefore, ZnO (that is the residue) will yield ng of ZnCl$_2$

Solve for n,

n = 8.1 x 136/81 = 13.6g

The mass of zinc chloride obtained = 13.6g

23. An excess of a divalent metal, M was dissolved in a limited volume of hydrochloric acid. If 576 cm^3 of hydrogen were liberated at s.t.p. what was the mass of the metal that produced this volume of hydrogen?

[M = 24, H = 1, molar volume of gases = 22.4dm^3]

Solution

> *Metals ionize as follows;*
> **Monovalent**
> M(s) → M$^+$ (aq) + 1 electron
> Na(s) → Na$^+$ (aq) + 1 electron
> 2Na(s) + 2HCl (aq) → 2NaCl (aq) H$_2$ (g)
>
> **Divalent**
> M(s) → M^{2+} (aq) + 2 electron
> Mg(s) → Mg^{2+} (aq) + 2 electron
> Mg(s) + 2HCl (aq) → MgCl$_2$ (aq) H$_2$ (g)
>
> **Trivalent**
> M(s) → M^{3+} (aq) + 3 electron
> Al(s) → Al^{3+} (aq) + 3 electron
> Al(s) + 3HCl (aq) → AlCl$_3$ (aq) 3H$_2$ (g)

Since the metal is divalent, it will react as follows;

M(s) + 2HCl (aq) → MCl$_2$ (aq) + H$_2$ (g)

(Remember that divalent elements carry a charge of 2)

We are concerned with the volume of hydrogen and the mass of the metal.

From the equation 1 mole of the metal liberates 1 mole of hydrogen gas.

So, 24g of the metal liberates 22.4dm^3 of hydrogen,

Therefore, ng of the metal will liberate 0.576dm^3 of hydrogen

Solve for n,

n = 24 x 0.567/22.4 = 0.62g

0.62g of the metal, M liberates 576cm^3 of hydrogen

24. 19.04g of ammonia gas were mixed with 31.10g of hydrogen chloride gas in a closed container.

a. Which of the reactants was in excess and by how much?
b. How much ammonium chloride was formed?
c. How much more of the insufficient reactant would be needed to completely react with the excess of the other reactant?

[N = 14, H=1, Cl= 35.5]

Solution
First, we write the equation for the reaction,
$NH_3 (g) + HCl(g) \rightarrow NH_4Cl(s)$
From the equation 1 mole of ammonia reacts with 1 mole of HCl
[Molar mass of ammonia = 17g, HCl = 36.5g]
Hence, 17g of ammonia normally reacts with 36.5g of HCl
So they react in the ratio 17:36.5, judging from this ratio, the reactant that was in excess was ammonia. To find by how much it was in excess, we calculate the amount of ammonia required to completely react with 31.10g of HCl. From the equation,
36.5g of HCl requires 17g of ammonia for complete reaction.
Therefore, 31.10g of HCl will require ng of ammonia for complete reaction.
Solve for n,

n = 31.10 x 17/36.5 = 14.48g

Only 14.48g of ammonia reacts with 31.10g of HCl,

Therefore, the mass of excess ammonia = 19.04 – 14.48 = 4.56g

b. From the equation of the reaction,

[Molar mass of ammonium chloride = 53.5g]

36.5g of HCl yields 53.5g of ammonium chloride.

Therefore, 31.10g of HCl will yield ng of ammonium chloride.

Solve for n,

n = 31.10 x 53.5/36.5 = 45.58g

The mass of ammonium chloride formed = 45.58g

We know that the mass of excess ammonia is 4.5g, to find how much of HCl, which will be, required to react with this excess.

From the equation,

17g of ammonia reacts with 36.5g of HCl.

Therefore, 4.56g of ammonia will react with ng of HCl.

Solve for n,

n = 4.56 x 36.5/17 = 9.79g

Therefore, 9.79g more of HCl will be required to react with the excess ammonia.

25. 20g of copper (ii) oxide was warmed with 0.05 mole of H_2SO_4 acid.

Calculate the mass of copper (ii) oxide that was in excess.
The equation for the reaction, $CuO(s) + H_2SO_4(aq) \rightarrow CuSO_4(aq) + H_2O(l)$
[O=16, Cu=64]

Solution

First we find the amount of CuO that would react with 0.05 mole of H_2SO_4 then the excess.

From the equation, 1 mole of H_2SO_4 acid is required to react with 1 mole of CuO.

[Molar mass of CuO= 80]

1 mole of H_2SO_4 requires 80g of CuO,

Therefore, 0.05 mole of H_2SO_4 will require ng of CuO

Solve for n,

n = 0.05 x 80/1 = 4

Therefore, 0.05 mole of H_2SO_4 acid requires 4g of CuO.

Then the excess CuO = 20 – 4 = 16g.

QUESTIONS

1. Calculate the volume of oxygen that was in excess if 150 cm^3 of carbon (ii) oxide was burnt in 80 cm^3 of oxygen according to the following equation,

    $2CO(g) + O_2 \rightarrow 2CO_2(g)$

2. What is the percentage by mass of sulphur in $Al_2(SO_4)_3$?

VOLUMETRIC ANALYSIS

A standard solution is a solution of known concentration.

The concentration of a solution is the amount of solute in a given volume of the solution. It is measured in either grams per dm^3 ($g\ dm^{-3}$) or mol per dm^3 ($mol dm^{-3}$).

The molar concentration of a solution is its concentration in $mol dm^{-3}$.

A molar solution of a compound is one which contains one mole or the molar mass of the compound in one dm^3 of the solution. For instance, a one molar solution of NaOH is one which contains 1 mole of NaOH or the molar mass of NaOH that is 40g of NaOH in one dm^3 of the solution.

So in order to prepare a one molar solution of NaOH, we dissolve 40g of NaOH in (and make it up to) one dm^3 of water.

DERIVATION OF MOLAR CONCENTRATION AND MASS CONCENTRATION

Substance	Molar mass(g)	Molar concentration(mol dm^{-3})	Mass concentration (g dm^{-3})
NaOH	40	1	40
NaOH	40	0.1	4
Na_2CO_3	106	1	106
Na_2CO_3	106	0.1	10.6
Na_2CO_3	106	0.5	53
H_2SO_4	98	1	98
H_2SO_4	98	0.1	9.8

EXERCISES

1. Calculate

(i) The mass of anhydrous sodium hydroxide, NaOH, present in 200 cm^3 of 0.1mol dm^{-3}

(ii) The number of NaOH particles present in the solution in (i) above. [Na=23, C=12, O=16]

Solution

(i) When presented with a problem like this we first calculate the mass of NaOH in 1 dm^3 of the solution.

The concentration of the solution is 0.1mol dm^{-3}

> Concentration in g dm^{-3} = Molarity x Molar

Using, Concentration in g dm^{-3} = Molarity x Molar mass

The molar mass of NaOH = 40g

Therefore, concentration in g dm^{-3} = 0.1 x 40 = 4.0g dm^{-3}

Since, 1 dm^3 of the solution contains 4g of NaOH

That is to say, 1000 cm^3 contains 4g of NaOH

200 cm^3 will contain n g of NaOH

> 1 dm^3 = 1000 cm^3

Cross multiply and solve for n,

n = (200 x 4)/1000 = 0.8g of NaOH

200 cm³ of the solution contains 0.8g of NaOH

(ii) To calculate the number of particles in the solution.

Number of NaOH particles, N = amount (moles) x 6.02 x 10^{23}

N = 0.1 x 6.02 x 10^{23} = 0. 602 x 10^{23}

This means that 1000 cm³ (1dm³) of 0.1M solution contains

0. 602 x 10^{23} Particles of NaOH.

Therefore, 300 cm³ of 0.1mol dm^{-3} solution will contain n particles

Cross multiply and solve for n,

n = $\dfrac{300 \times 0.602 \times 10^{23}}{1000}$ = 0.181 x 10^{23} Particles of NaOH

2. Calculate the volume of 0.15mol dm^{-3} solution of hydrochloric acid, HCl, that will contain a mass of 3g of the raw acid [H=1, Cl=35.5]

Solution

The molar mass of HCl = 36.5g

The mass of raw HCl in 1 dm^3 of the 0.15M solution, which is the concentration, can be calculated as follows.

The concentration in gdm^{-3} of the solution = Molarity x Molar mass

The concentration in gdm^{-3} of the HCl solution = 0.15 x 36.5 = 5.5g dm^{-3}

That is 5.5g of HCl is contained in 1 dm^3 of 0.15M solution.

Therefore, 3g of HCl will be contained in n dm^3 of 0.15M solution

Cross multiply, solve for n,

n =(3 x 1)/5.5 = 0.55 dm^3

Therefore, 0.55 dm^3 of 0.15M HCl solution will contain 3g of the raw HCl acid.

3. Calculate the volume of hydrogen chloride gas at s.t.p. that would yield 2dm^3 of 0.25mol dm^{-3} aqueous hydrogen chloride solution.

[Molar volume of gases = 22.4 dm^3]

Solution

First we calculate the number of mol contained in 2 dm^{-3} of the solution.

The concentration of HCl solution required = 0.25mol dm^{-3}

1 dm^3 of the HCl solution contains 0.25mol.

Therefore, 2 dm^3 of the HCl solution will contain n mol.

Solve for n,

n = (2 x 0.25)/1 = 0.5mol

2 dm^3 of the HCl solution contains 0.5 mol.

Next we calculate the volume occupied by 0.5 Mol

We know that, 1 mole of HCl gas occupies 22.4 dm^3 at s.t.p., therefore, 0.5 mole of HCl gas will occupy n dm^3.

Solve for n,

n = 0.5 x 22.4/1 = 11.2 dm^3.

So, 11.2 dm^3 of hydrogen chloride gas at s.t.p will be required to yield

2dm^3 of a 0.25mol dm^{-3} aqueous solution.

DILUTION OF SOLUTIONS

Dilution is the addition of a definite volume of a solvent, to a definite volume of a concentrated solution (solution). During this dilution the number of moles present in the concentrated solution does not change. It's a bit like placing 3 cubes of sugar in a glass and then adding water till the glass is quarter full. If you later add more water till the glass is filled to the brim, you still have the same quantity of sugar, but more sugar solution.

Number of mole of a solute (amount) = molar concentration (C) x volume (v) in dm^3.

During the dilution of a solution, the product of its initial volume and concentration is equal to the product of the final volume and concentration.

$$C_1V_1 = C_2V_2$$
C1 = initial molar concentration,
C2 = final molar concentration,
V1 = initial volume of concentrated solution,
V2 = final volume of solution.

That is, $C_1V_1 = C_2V_2$.

Where, C1 = initial molar concentration,

C_2 = final molar concentration,

V_1 = initial volume of concentrated solution,

V_2 = final volume of solution.

For instance, if 300 cm^3 of distilled water is added to 100 cm^3 of a 2M NaOH solution, the solution is diluted to four times its initial volume, hence its concentration will decrease by four times its initial concentration. That is, the volume of the solution varies inversely with its concentration. As you increase the volume the concentration decreases, and as you decrease the volume the concentration increases.

Exercises

1. Calculate the final molar concentration of NaOH. When 100 cm^3 of distilled water is added to 50 cm^3 of a 3M NaOH solution to produce 150 cm^3 of the diluted solution.

Solution

Using, $C_1V_1 = C_2V_2$

C_1 = 3M, V_1 = 50 cm^3, V_2 = 150 cm^3, C_2 is unknown

$C_2 = (C_1 \times V_1)/V_2 = (3 \times 50)/150 = 1M$.

The final molar concentration = 1 M

2. How much distilled water must be added to 70 cm^3 of a 2M NaOH solution in order to get a 0.5M NaOH solution.

Solution

First we have to find the new volume of the solution, which will contain 0.5M NaOH. Using, $C_1V_1 = C_2V_2$

$C_1 = 2M$, $V_1 = 70$ cm^3, $C_2 = 0.5M$, V_2 is unknown.

If we substitute in the equation, then final volume will be,

$V_2 = (2 \times 70)/0.5 = 280$ cm^3

The new volume to which it must be diluted = 280 cm^3

Therefore, **the amount of distilled water to be added = 280 − 70 = 210 cm^3**

3. A 1M tetraoxosulphate (vi) solution was diluted, what volume of it must be made up to 250 cm^3 to make it 10 times diluted. Hence, calculate the volume of distilled water used.

Solution

First we have to get the final concentration of the solution.

Since, the 1M solution is diluted to ten times its concentration. The new concentration will be 1/10 = 0.1M.
Now using, $C_1V_1 = C_2V_2$, We can find the new volume.
$C_1 = 1M$, $C_2 = 0.1M$, $V_1 = ?$, $V_2 = 250$ cm^3

$V_1 = (0.1 \times 250)/1 = 25$ cm^3.

So 25 cm^3 must be made up to 250 cm^3 to dilute it to 10 times its concentration.

Therefore the volume of distilled water used = 250 – 25 = 225 cm^3

Alternatively, you could treat this question with common sense.

If it is diluted 10 times that means the volume was increased 10 times to make it 250 cm^3. That means the starting volume is a tenth of 250, which is 25 cm^3.

CALCULATIONS USING TITRATION RESULTS

Candidates may be required to determine the following from the results of their titration of acid and bases, or at times they may be given the necessary values and required to calculate the following.

1. The molar concentration of an acid or base in moldm^{-3}.

2. The mass concentration of an acid or base in g dm^{-3}.

3. Molar mass of acids/bases.

4. Percentage purity/impurity of acids or bases.

5. Percentage of water of crystallization.

6. Solubility of acids or bases in moldm^{-3} or g dm^{-3}.

7. Mole ratio of reacting acids or bases.

Now let us solve some simple problems to enable us further our understanding of calculations from results of experiments.

Exercises

1. 25 cm^3 of a solution of NaOH containing 6g of the alkali per dm^3 required 18.0 cm^3 of a tetraoxosulphate (vi) solution for complete neutralization. Calculate, (i) The molar concentration of the NaOH solution. (ii) The concentration of the tetraoxosulphate (vi) solution.

Solution

(i) We can calculate the molar concentration of NaOH using, Molar concentration = concentration in g dm^{-3}/molar mass.

Since the mass concentration is given as 6g of NaOH per dm^3 and the molar mass of NaOH = 23+16+1=40g

Molar concentration = 6/40 = 0.15mol dm^{-3}.

The molar concentration of the NaOH solution = 0.15mol dm^{-3}.

(ii) Since we are now armed with the molarity of NaOH, the volume of the acid and base, we can now calculate the concentration of tetraoxosulphate (vi) solution using the formula;

$C_A V_A / C_B V_B = N_A / N_B$

> $(C_A V_A)/(C_B V_B) = N_A/N_B$
> Where
> C_A = Molar concentration of acid,
> C_B = Molar concentration of base,
> V_A = Volume of acid,
> V_B = Volume of base,
> N_A = Number of moles of acid,
> N_B = Number of moles of base.

Where,

C_A = molar concentration of acid,

C_B = molar concentration of base,

V_A = volume of acid, V_B = volume of base,
N_A = number of moles of acid,

N_B = number of moles of base.

Equation for the reaction,

$2NaOH(aq) + H_2SO_4(aq) \rightarrow Na_2SO_4(aq) + H_2O(l)$
from the equation the mole ratio of acid to base is 1:2.

substituting our values in $C_A V_A / C_B V_B = N_A / N_B$

we have

$(C_A \times 18)/(0.15 \times 25) = \frac{1}{2}$

$C_A = 0.104 moldm^{-3}$. So the molar concentration of the tetraoxosulphate (vi) solution = $0.104 moldm^{-3}$
The mass concentration of H_2SO_4 can be found using,

Mass concentration = molar concentration x molar mass

Mass concentration of H_2SO_4 = 0.104 x 98 = 10.2g dm^{-3}

2. 25 cm^3 of an 0.1mol dm^{-3} tetraoxosulphate(vi)solution neutralized 19.0 cm^3 of a solution containing 3.0g of impure NaOH in 250 cm^3 of solution. Calculate the percentage purity of the NaOH sample.

Solution

The equation for the reaction
$2NaOH(aq) + H_2SO_4(aq) \rightarrow Na_2SO_4(aq) + H_2O(l)$
from the equation 1 mole of the H_2SO_4 neutralized 2 moles of NaOH. First, we have to find the amount (concentration) of pure NaOH which reacted with $25cm^3$ of $0.1 moldm^{-3}$ H_2SO_4

using $C_AV_A/C_BV_B = N_A/N_B$,
(0.1 x 25)/ (C_B x 20) = ½ ,
C_B = 0.1 x 25 x 2/20 x 1 = 0.25mol dm^{-3}.
Now let us convert the concentration of pure NaOH which reacted to gdm^{-3}.
Using, mass concentration = molar concentration x molar mass,

mass concentration of NaOH = 0.25 x 40 = 10gdm^{-3}.
Therefore the mass of pure NaOH which reacted is 10gdm^{-3}.

In the question the mass of impure NaOH is given as 3.0g per $250cm^3$.

Since 250cm³ of the solution contains 3.0g of impure NaOH

1000cm³ of the solution will contain n g of impure NaOH

we cross multiply and solve for n,

n = 1000 x 3.0/250 = 12g dm^{-3}.

% purity of a sample
= [mass of the pure sample/mass of impure sample] x 100

% impurity
= [mass of impure – mass of pure sample/mass of impure] x 100

% purity of a sample

= [mass of the pure sample/mass of impure sample] x 100

= [10g dm^{-3}/12g dm^{-3}] x 100 = 83.33%

we could also calculate the percentage impurity if required.

% impurity = [mass of impure – mass of pure sample/mass of impure] x 100

= [12 – 10/12] x 100 = 16.66%

3. 29.00cm³ of a solution containing 0.095mol dm^{-3} of HCl solution neutralized 25cm³ of a solution containing 15.76g of $Na_2CO_3.xH_2O$. Calculate (i) the concentration of B in mol dm-³, (ii) the value of x.

Solution

The equation for the reaction,

$2HCl(aq) + Na_2CO_3 \cdot xH_2O(aq) \rightarrow 2NaCl(aq) + (x+1)H_2O(l)$.

From the equation the mole ratio of the acid to base is 2:1

(i) We can calculate the concentration of B in mol dm^{-3} using

$C_A V_A / C_B V_B = N_A / N_B$

Substituting, $0.095 \times 29 / C_B \times 25 = 2/1$,

$C_B = 0.095 \times 29 \times 1/25 \times 2 = 0.055$ mol dm^{-3},

the concentration of the base that reacted with the acid = 0.055 mol dm^{-3}.

(ii) To solve for the value of x,

the molar mass of $Na_2CO_3 \cdot xH_2O$

$= [(2 \times 13)+(1 \times 12)+(3 \times 16)] + x[(1 \times 2)+(1 \times 16)] = 106 + 18x$.

We can determine the value of x by using,

molar concentration = mass concentration/molar mass. In the equation the mass concentration of $Na_2CO_3 \cdot xH_2O$ is given as 15.76g dm^{-3} and we found the molar concentration to be 0.055 mol dm^{-3}.

So $0.055 = 15.76/106 + 18x$

$106 + 18x(0.055) = 15.76$

$5.83 + 0.99x = 15.76$

$0.99x = 15.76 - 5.83$

$0.99x = 9.93$

$x = 9.93/0.99$

x = 10.03

So the value of x is approximately 10, hence we have $Na_2CO_3.10H_2O$

4. 28.0 dm^{-3} hydrochloric acid, HCl, of concentration 4.1g dm^{-3} neutralized 25.0cm^3 of an unknown alkali, YOH, whose concentration was 7.0gdm^{-3}. Calculate the (a) the molarity of YOH (b) the relative atomic mass of the element Y. Name the element if possible. (HCl = 36.5)

Solution

Equation for the reaction, HCl + YOH → YCl + H_2O
from the equation one mole of the HCl neutralizes 1 mole of YOH

(a) The molarity of HCl can be determined using,

Concentration in g dm^{-3} = Molarity x Molar mass

Molarity of HCl = concentration in gdm^{-3}/molar mass.
= 4.1/36.5 = 0.112mol dm-^{3}.

Knowing the molarity of HCl and the volume of the base used we can now calculate the molarity of YOH,
using, $C_A V_A / C_B V_B = N_A / N_B$,
substituting, 0.112 x 28 x 1/C_B x 25 = 1/1,
C_B = 0.112 x 28/25 = 0.125mol dm^{-3}.
The molarity of YOH = 0.125mol dm^{-3}.

(b) Let y be the relative atomic mass of Y.

Molar mass of YOH = y + 16 + 1 = y + 17.

Applying molar concentration = mass concentration/molar mass.

Remember, the mass concentration of YOH is given as 7.0g dm^{-3}.

Substituting, 0.125 = 7/y + 17,

0.125 (y + 17) = 7,

0.125y + 2.132 = 7,

0.125y = 7 – 2.132,

0.125y = 4.868,

y = 4.868/0.125 = 38.9

The relative atomic mass of Y is approximately = 39, thus Y is potassium.

5. Solution B was made by taking 20cm³ of a saturated solution of sodium trioxocarbonate (iv) at 25 °C and diluting it to 1000cm³ with distilled water. If 25cm³ of solution B neutralized 24.40cm³ of a 0.090moldm⁻³ hydrochloric acid. Calculate, (a) The concentration of B in moldm³. (b) The solubility of sodium trioxocarbonate (iv) at 25°C in moldm⁻³ (c) The mass of sodium trioxocarbonate (iv) that would be obtained by evaporating 1.0dm³ of the saturated solution to dryness.

Solution

(a) we can calculate the concentration of solution B in $moldm^{-3}$ by using, $C_A V_A / C_B V_B = N_A / N_B$,

equation for the reaction,

$2HCl(aq) + Na_2CO_3(aq) \rightarrow 2NaCl(aq) + H_2O(l)$

from the equation the mole ratio of the acid to base is 2:1.

$C_A = 0.090 moldm^{-3}$, $C_B = ?$, $V_A = 24.40 cm^3$, $V_B = 25 cm^3$, $N_A = 2$, $N_B = 1$

$C_B = C_A V_A N_B / V_B N_A = 0.090 \times 24.4 \times 1/25 \times 2$,

$C_B = 0.044 moldm^{-3}$.

The concentration of solution B = $0.044 moldm^{-3}$.

(b) The solubility of sodium trioxocarbonate (iv) at 25 °C is the same as the concentration of the saturated solution at 25 °C. We can calculate the concentration of the saturated solution which was diluted to get the new solution, using, $C_1 V_1 = C_2 V_2$,

Where C_1 = initial concentration = ?,

C_2 = final concentration = $0.044 mol\ dm^{-3}$

V_1 = initial volume = $20 cm^3$,

V_2 = final volume = $1000\ cm^3$,

$C_1 = C_2 V_2 / V_1 = 0.044 \times 1000/21 = 2.095 mol\ dm^{-3}$.

Therefore, the solubility of sodium trioxocarbonate(iv) at 25 °C = $2.095 mol\ dm^{-3}$.

(c) To calculate the mass of sodium trioxocarbonate (iv) that would be obtained on evaporating $1.0\ dm^{-3}$ of the saturated solution. We have to find the mass of sodium trioxocarbonate

(iv) in 1.0 dm^{-3}, and this would give us the mass obtained on evaporation.

Using, mass concentration = molar concentration x molar mass.

Molar mass of sodium trioxocarbonate (iv)
= (23x2) + 12 + (16x3) = 106. Concentration in g dm^{-3} = 2.095 x 106 = 222.1g dm^{-3}.

So the mass of sodium trioxocarbonate(iv) that would be obtained on evaporating 1.0 dm^3 of the saturated solution = 222.1g

6. 17.30cm^3 of a solution containing 3.70gdm^{-3} of an acid H_2X neutralized 25cm^3 of a solution containing 2.15g of sodium hydroxide per 250cm^3 of solution.

Calculate, (a) concentration of solution B in mol dm^{-3}, (b) concentration of solution A in mol dm^{-3}, (c) molar mass of the acid, H_2X

Solution

(a) To calculate the concentration of B in moldm^{-3}. B is a solution containing 2.15g of sodium hydroxide per 250cm^3.
Since, 250cm^3 of solution B contains 2.15g of NaOH
1000cm^3 of solution B will contain ng of NaOH, solve for n,
n = 1000 x 2.15/250 = 8.6g
Hence, the concentration of B in gdm^{-3} = 8.6gdm^{-3}.

Remember, mass concentration = molar concentration x molar mass.

Molar concentration of B = 8.6/40 = 0.215 moldm^{-3}.

(b) We can calculate the concentration of A in moldm^{-3},

Using, $C_A V_A / C_B V_B = N_A / N_B$

The equation for the reaction,

$H_2X(aq) + 2NaOH(aq) \rightarrow Na_2X(aq) + H_2O(l)$

From the equation 2moles of NaOH reacts with 1mole of the acid, H_2X.

$C_A = C_B V_B N_A / V_A N_B$

$C_B = 0.215$ moldm^{-3}

$V_A = 17.30$ cm^3

$V_B = 25$ cm^3

$\quad N_A = 1$

$\quad N_B = 2$

$\quad C_A = \underline{0.215 \times 2.5 \times 1}$
$\quad\quad\quad\quad 17.80 \times 2 \quad = 0.155$ moldm^{-3}

The concentration of A = 0.155 moldm^{-3}

(c) To find the molar mass of the acid, H_2X

Remember,

Concentration in gdm^{-3} = molar concentration x molar mass

Hence, molar mass = $\underline{\text{molar concentration in gdm}^{-3}}$
$\quad\quad\quad\quad\quad\quad\quad\quad\quad$ Molar concentration

Concentration of H_2X in gdm^{-3} = 3.70 gdm^{-3}

Molar concentration of H_2X = $\underline{3.70}$
$\quad\quad\quad\quad\quad\quad\quad\quad\quad\quad\quad$ 0.155

=23.9g

The molar mass of H_2X is approximately = 24g

7. 25 cm³ of an 0.09 moldm⁻³ solution of NaOH is neutralized by 22.6 cm³ of an acid solution containing 4.5g of the acid per gdm⁻³ of solution. Deduce the mole ratio of acid to Alkali given that the molar mass of the acid is 98g

SOLUTION

First we determine the molar concentration of the acid, using

Molar concentration = $\dfrac{\text{concentration in gdm}^{-3}}{\text{Molar mass}}$

$= \dfrac{4.5}{90} = 0.05 \text{ moldm}^{-3}$

Now using

$C_A V_A / C_B V_B = N_A / N_B$

Where N_A/N_B = mole ratio of acid to alkali

$C_A = 0.05 \text{ moldm}^{-3}$

$C_B = 0.09 \text{ moldm}^{-3}$

$V_A = 22.60 \text{ cm}^3$

$V_B = 25 \text{ cm}^3$

Substituting,

$\dfrac{0.05 \times 22.6}{0.09 \times 25} = \dfrac{\text{mole of acid}}{\text{mole of base}}$

Mole of acid   = 1.13
Mole of base   2.25

Simplifying we have,

Mole of acid   = 1
Mole of base   2

Therefore, mole ratio of acid to base = 1:2

8. A piece of zinc was added to 1000 cm^3 of 0.2M hydrochloric acid. After effervescence had stopped, 28 cm^3 of the resulting solution required 17 cm^3 of 0.08M sodium trioxocarbonate (iv) solution, for complete neutralization. Calculate the mass of the zinc added. (Zn=65, HCL= 36.5, Na_2Co_3 = 106)

SOLUTION

First, we need to find the actual concentration of HCL acid, which neutralized the Na_2Co_3. Using

$C_A V_A / C_B V_B = N_A / N_B$

C_A = ?
C_B = 0.08M
V_A = 28 cm^3
V_B = 17 cm^3
N_A = 2
N_B = 1

$$C_A = \frac{C_B V_B N_A}{V_A N_B}$$

$$C_A = \frac{0.08 \times 17 \times 2}{28 \times 1} \quad 0.097 \text{ moldm}^{-3}$$

The actual concentration of HCl, which was left after reacting with Na_2Co_3

$= 0.097$ moldm^{-3}

Hence, the amount of the HCl left after reacting with the Zinc is 0.097M

The amount of HCl which reacted with the Zinc = 0.2-0.097 = 0.103M

The equation for the reaction

$2HCl(aq) + Zn(s) \rightarrow ZnCl_2(aq) + H_2(g)$

From the equation, 2 moles of HCl reacts with 1mole of Zinc

Therefore, 0.103mole of HCl will react with x mole of Zinc

Cross multiply and solve for x

$X = \frac{0.103 \times 1}{2} = 0.05$ Mole

The amount of Zinc added was 0.05Mole. We can convert this quantity to mass using,

No of mole = Reacting mass /Molar mass

Therefore, reacting Mass of Zinc = no of mole x molar mass

$= 0.05 \times 65$

$= 3.34$g

So the mass of zinc added was 3.34g.

9. 2g of a mixture of sodium hydroxide and sodium chloride (as impurity) were dissolved in 500 cm³ of water. If 2.5 cm³ of the solution were neutralized by 21.0 cm³ of 0.1M hydrochloric acid, calculate the percentage of the sodium chloride impurity. (NaOH = 40, HCl = 36.5, NaCl = 58.5)

SOLUTION

First let us find the amount of NaOH, which was in the mixture. That is the amount of NaOH, which reacted with the HCl, since the NaCl doesn't react with the acid.

Using

$C_A V_A / C_B V_B = N_A / N_B$

NaOH(aq) + HCl(aq) → NaCl(aq) + H$_2$O(l)

The mole ratio of acid to base = 1:1

$$C_B = \frac{C_A V_A \; N_B}{V_B \; N_A}$$

C_A = 0.1M

C_B = ?

V_A = 21.0 cm³

V_B = 25 cm³

Substituting,

$$\frac{0.1 \times 21.0 \times 1}{25 \times 1} = 0.084 \; moldm^{-3}$$

So the amount of NaOH present in the mixture can be found thus,

1000cm³ of the solution contains 0.084 mole of NaOH

∴500cm of the solution will contain x mole of NaOH

Cross multiply and solve for x

X = $\frac{500 \times 0.084}{1000}$ = 0.042

The actual amount of NaOH in the mixture was 0.42 mole. Let us convert this to mass. Remember

Reacting mass = no of mole x molar mass

= 0.042 x 40

=1.68g

So the mass of NaOH in the mixture was 1.68g.

∴ The mass of NaCl = 2 - 1.68 = 0.32g

Hence the percentage of NaCl impurity = $\frac{0.32 \times 100}{2}$ = 16%

10. Calculate the mass of pure sodium chloride that will yield enough hydrogen chloride gas to neutralize 25cm³ of 0.5m potassium trioxocarbonate (IV) solution

(NaCl = 58.5, HCl = 36.5, K_2CO_3 = 138)

Solution

First let us find the amount of the K_2CO_3 in 25cm³ of 0.5M, then the amount of HCl gas that would react completely with it.

1000cm³ of the solution contains 0.5mole of K_2CO_3

∴25cm³ of the solution will contain x mole of K_2CO_3

Cross multiply and solve for x

$$X = \frac{25 \times 0.5}{1000} = 0.0125 \text{ mole of } K_2CO_3$$

$2HCl(l) + K_2CO_3(aq) \rightarrow 2KCl(aq) + CO_2(g) + H_2(g)$

From the equation

2mole of HCl reacts completely with 1mole of K_2CO_3

∴ x mole of HCl will react with 0.0125 mole of K_2CO_3

Cross multiply and solve for x

$$X = \frac{2 \times 0.0125}{1} = 0.025 \text{ mole}$$

So the amount of HCl gas required, to neutralize 25cm^3 of 0.5M K_2CO_3 is 0.025 mole. Now lets find the mass of NaCl, which will yield 0.025 mole of HCl gas.

The equation for the reaction

$2NaCl(s) + H_2SO_4(aq) \rightarrow 2HCl(g) + Na_2SO_4(aq)$

from the equation 2 moles of NaCl yields 2 moles of HCl.

That is (2 x 58.5)g of NaCl yields 2 moles of HCl gas

∴ Xg of NaCl will yield 0.25mole of HCl gas

Cross multiply and solve for x

$$x = \frac{2 \times 58.5 \times 0.025}{2}$$

x = 1 . 46g

1.46g of NaCl will yield enough HCl gas to neutralize 25cm³ of 0.5M K_2CO_3 solution.

11. 20.20cm³ of a solution of H_2SO_4 acid neutralizes 25cm³ of a solution containing 2.8g of potassium hydroxide per 250cm³
Calculate
(a) The concentration of the base in moldm^{-3}
(b) The concentration of the acid in moldm^{-3}
(c) The number of hydrogen ions in 1.0dm3 of acid .

Equation for the reaction
H_2SO_4 (aq) + 2KOH(aq) → K_2SO_4 (aq) + $2H_2O$ (l)
(H = 1, O = 16, K = 39, Avogadro's constant = 6.02×10^{23})

Solution

(a) To calculate the concentration of the KOH solution in moldm^{-3} , first, we find the mass of KOH contained in 1dm³ of the solution

250 cm3 of the solution contains 2.8g KOH
1000cm3 of the solution will contain xg of KOH
cross multiply and solve for x

x = $\underline{1000 \times 2.8}$
　　　250　　　　= 11.2g

The concentration of the KOH solution = 11.2gdm^{-3}
We convert this to moldm^{-3},

Using, concentration in $moldm^{-3}$ = concentration in glm^{-3}/molar mass

Molar mass of KOH = 39+16 + 1 = 56 molar mass

Concentration in moldm-3 of KOH = $\dfrac{11.2}{56}$ = 0.2 $moldm^{-3}$

Concentration of KOH solution = 0.2 $moldm^{-3}$

(b) The concentration of the H_2SO_4 acid in $moldm^{-3}$

$C_A V_A / C_B V_B = N_A / N_B$

C_A = ?, C_B = 0.2 $moldm^{-3}$, V_A = 20.2, V_B = 25 cm^3, N_A = 1, N_B = 2

$C_B = \dfrac{C_A V_A N_B}{V_B N_A}$

$C_A = \dfrac{0.2 \times 25 \times 1}{20.2 \times 2}$ = 0.124 $moldm^{-3}$

(c) The number of hydrogen ions in $1 dm^3$ of the H_2SO_4 acid solution.

Ionisation equation

$H_2SO_4 \rightarrow 2H^+ + SO_4^{2-}$

From the equation 1 mole of the acid yields 2 moles of hydrogen ion

Remember 1 mole of hydrogen ion contains 6.02×10^{23} ion

So 1 mole of H_2SO_4 contains $2 \times 6.02 \times 10^{23}$ hydrogen ion

∴ 0.124 mol of H_2SO_4 will contain x hydrogen ion

Cross multiply and solve for x

$$x = \frac{0.124 \times 2 \times 6.02 \times 10^{23}}{1} = 1.49 \times 10^{23} \text{ hydrogen ion}$$

The number of hydrogen ions in $1dm^3$ of the H_2SO_4 solution
= 1.49×10^{23} hydrogen ions.

12. What volume of $0.1 moldm^{-3}$ solution of tetraoxosulphate (iv) acid would be needed to dissolve 2.86g of sodium trioxocarbonate (iv) decahydrate crystals.
(H=1, C=12, O=16, S=32, Na=23)

Solution

Equation for the reaction

$H_2SO_4(aq) + Na_2CO_3(aq) \rightarrow 2NaCl(aq) + H_2O(l) + CO_2(g)$

from the equation 1 mole of H_2SO_4 reacts with 1 mole of Na_2CO_3

That is 1 mole H_2SO_4 reacts with 106g of Na_2CO_3

∴ x mole of H_2SO_4 will react with 2.86g of Na_2CO_3

cross multiply and solve for x

$$X = \frac{1 \times 2.86}{106} = 0.027 mole$$

0.027 mole of H_2SO_4 will react with 2.86g of Na_2CO_3

we know from the concentration of H_2SO_4 that,

$1000cm^3$ contains 0.1 mole of H_2SO_4

13. 200cm³ each of 0.1m solutions of lead (ii) trioxonitrate (v) and hydrochloric acid were mixed. Assuming that lead (ii) chloride is completely insoluble, calculate the mass of lead (ii) chloride that will be precipitated.
(Cl = 35.5, N = 14, O = 16, Pb = 207)

Solution

Equation of the reaction

$Pb(NO_3)_2$ (aq) + $2HCl$(aq) → $PbCl_2$ (aq) + $2HNO_3$(aq)

From the equation 1 mole of $Pb(NO_3)_2$ reacts with 2 moles of HCl to form 1 mole of $PbCl_2$

Now let us find the amount of HCl in 200cm³ of 0.1M solution,

From the concentration 1000cm³ contains o.1mole of HCl

∴ 200cm³ will contain x mole of HCl

Cross multiply and solve for x

$$X = \frac{200 \times 0.1}{1000} = 0.02 \text{ mole of HCl}$$

We know that

2 moles of HCl form 1 mole of $PbCl^2$

that is, 2 moles of HCl form 278g of $PbCl_2$

∴ 0.02 moles of HCl form xg of $PbCl_2$

Solve for x

$$X = \frac{0.02 \times 278}{2} = 2.78g$$

2.78g of $PbCl_2$ will be precipitated.

14. Mg (s) + 2HCl(aq) → $MgCl_2$ (aq) + H_2 (g)

From the equation above, the mass of magnesium required to react with $250cm^3$ of 0.5M HCl is ? (Mg = 24)

Solution

First we find the amount of HCl in $250cm^3$ of 0.5M solution

$1000cm^3$ contains 0.5 mole of HCl

∴ $250cm^3$ will contain x mole of HCl

Solve for x

X = $\underline{250 \times 0.5}$
         1000   = 0.125 mole of HCl

From the Equation

I mole of Mg reacts with 2 mole of HCl

that is 24g of Mg reacts with 2 mole of HCl

∴ xg of mg will react with 0.125 mole of HCl

Solving for x

X = $\underline{24 \times 0.125}$
         2   = 1.5g of magnesium

15. 0.25 mole of hydrogen chloride was dissolved in distilled water and the volume made up to $0.50dm^3$. If $15.00cm^3$ of the solution requires 12.50cm3 of aqueous sodium trioxocarbonate (iv) for neutralization, calculate the concentration of the alkaline solution.

Solution

First let us find the concentration of the hydrochloric acid. From the question $0.5dm^3$ of the solution contains 0.25mole of HCl

∴ 1 dm^3 off the solution will contain x mole HCl
Solving for x

$$X = \frac{1 \times 0.25}{0.5} = 0.5 \text{ mole}$$

∴ the concentration of the HCl solution = $0.5 moldm^{-3}$,
we can now find the concentration of the alkaline solution using

$C_AV_A/C_BV_B = N_A/N_B$

Equation for the reaction

2 HCl (aq) + Na_2CO_3 (aq) → 2 NaCl (aq) + H_2O (1)

From the equation mole ratio of acid to base is 2:1

$$C_B = \frac{C_AV_A N_B}{V_B N_A}$$

C_A = 0.5M, C_B = ? N_A = 2
V_A = $15.00cm^3$, V_B = $12.50cm^3$ N_B = 1
Substituting

$$C_B = \frac{0.5 \times 15 \times 1}{12.5 \times 2} = 0.3 moldm^{-3}$$

The concentration of the Alkaline solution $(Na_2CO_3) = 0.30$ moldm^{-3}.

16. 25cm^3 of a 0.2 moldm^{-3} solution of Na_2CO_3 requires 20cm^3 of a solution of HCl for neutralization. The concentration of the HCl solution is ?

Solution
Equation for the reaction
$2HCl\ (aq) + Na_2CO_3\ (aq) + H_2O\ (l)$
from the equation the mole ratio of acid to base is 2 :1
using , $C_AV_A/C_BV_B = N_A/N_B$

$$C_A = \frac{C_B V_B\ N_A}{V_A\ N_B}$$

$C_B = 0.2$moldm$_3$, $V_B = 25$cm3
$V_A = 20$cm^3, $N_A = 2$, $N_B = 1$

$$C_A = \frac{0.2\ \text{x}\ 25\ \text{x}\ 2}{20\ \text{x}\ 1} = 0.5 \text{moldm}^{-3}$$

The concentration of the HCl solution = 0.5moldm^{-3}

17. A hydrated salt of formula $MSO_4.x\ H_2\ O$
Contains 45.3% by mass of water of crystallization. Calculate the valve of x .
(M = 56, S = 32, O = 16, H = 1)

Solution

The molar mass of the hydrated salt

$MSO_4 \cdot XH_2O$ = 56 + 32 + (16x4) +X[(2x1)+16]

= 152 + 18X

The mass of the anhydrous salt MSO_4

= 56 + 32 + (16+4) + 152

Remember

% Water of crystallization

= Mass of hydrated - mass of anhydrous salt x 100

                 Mass of hydrated salt

Substituting

45 = 152 +18X - 152   x 100
        152 + 18X

Solve for X

45 =   18X   x 100
     152 + 18X

45 =   1800X
     152+ 18X

45 (152 +18X)   = 1800X

6840 + 810X = 1800X

6840 = 1800X − 810X

6540 = 990X

X=   6840
     990   = 6.9

The value of X is approximately = 7

18. 29.20cm^3 of a 0.05 moldm^{-3} hydrochloric acid neutralized 25cm^3 of 0.025 moldm^{-3} of a trioxocarbonate (iv) solution.

a. Calculate the mole ratio of trioxocabonate (iv) in the reaction.

b. Given that the alkali contains 7.2gdm^{-3} of the hydrated trioxocarbonate (iv) salt calculate.

    (i)   Concentration of the anhydrous salt in gdm^{-3}
          (Molar mass of the anhydrous salt = 106)

    (ii)   Percentage of water of crystallization in the hydrated salt.

Solution

a. Using, $C_A V_A / C_B V_B = N_A / N_B$

    Where N_A / N_B = mole ratio of acid to base

Substitution

$$\frac{0.05 \times 29.20}{0.025 \times 25} = \frac{N_A}{N_B}$$

$$\frac{1.46}{0.625} = \frac{N_A}{N_B}$$

Simplifying further,

$$\frac{2.3}{1} = \frac{N_A}{N_B}$$

Approximately mole ratio of acid to base is 2:1

(i) The concentration of the anhydrous salt is the same as the concentration of the salt which neutralized the acid.

The concentration of the salt is 0.025moldm^{-3}. The molar mass of the salt is given as 106g. We can calculate the concentration of the salt in gdm^{-3}, using

Concentration in gdm^{-3}

= concentration in moldm^{-3} x molar mass

= 0.025 x 106 = 2.65gdm^{-3}

(ii) The mass of the anhydrous salt = $2..65 \text{gdm}^{-3}$

Mass of hydrated salt is given as 7.2gdm^{-3}

∴ Mass of water of crystallization = 7.2 – 2.65

= 4.55gdm^{-3}

% Water of crystallization

= $\dfrac{\text{Mass of water of crystallization}}{\text{Mass of hydrated salt}}$ x 100

= $\dfrac{4.55}{7.2}$ x 100 = 63.2%

19. A solution of calcium bromide contains 20gdm^{-3}, what is the molarity of the solution with respect to calcium bromide and bromide ions?

Solution

Equation for the ionization of calcium bromide

$CaBr_2 \rightarrow Ca^{2+} + 2Br^-$

We can find the molarity of the solution with respect to $CaBr_2$ by

$$\text{Concentration in } moldm^{-3} = \frac{\text{Concentration in } gdm^{-3}}{\text{Molar mass}}$$

Molar mass of $CaBr_2 = 40 + (80 \times 2) = 200g$

Concentration $moldm^{-3}$ of $CaBr_2 = \frac{20}{200}$

$= 0.1 moldm^{-3}$

To find the concentration with respect to bromide ions from the ionization equation

1 mole of $CaBr_2$ yields 2 moles of Br^- on ionization.

∴ 0.1 mole of $CaBr_2$ will yield x mole of Br^- on ionization

Solving for x

$x = \frac{0.1 \times 2}{1} = 0.2 \; moldm^{-3}$

So the molarity of the solution with respect to calcium bromide, and bromide ion is $0.1 \; moldm^{-3}$ and $0.2 \; moldm^{-3}$ respectively.

20. On recrystallization, 20g of magnesium tetraoxosulphate (vi) forms 41g of magnesium tetraoxosulphate (vi) crystals, $MgSO_4.YH_2O$. The value of Y is?
($Mg = 24$, $S = 32$, $O = 16$, $H = 1$)

Solution

From the question we can deduce that the mass of the Anhydrous salt = 20g,

The mass of the Hydrated salt = 41g

∴ The mass of water of crystallization =

Mass of hydrated − mass of anhydrous salt

= 41 − 20 = 21g

The percentage by mass water of crystallization in the salt

= $\frac{21}{41}$ × 100 = 51.2%

The molar mass of the anhydrous salt,

$MgSO_4$ = 24 + 32 + (16 × 4) = 120g

The molar mass of the hydrated salt ,

$MgSO_4 \cdot YH_2O$ = 24 + 32 + (16 × 4) + Y[(2 × 1) + 16]

= 120 + 18y

Remember,

% Water of crystallization

= $\frac{\text{Mass of hydrated} - \text{Mass of anhydrous}}{\text{Mass of hydrated salt}}$ × 100

Substituting

51.2% = $\frac{120 + 18Y - 120}{120 + 18Y}$ × 100

51.2 = $\frac{(18Y) 100}{120 + 18Y}$

51.2 = $\frac{1800Y}{120 + 18Y}$

51.2(120 + 18Y) = 1800 Y

6144 + 921.6Y = 1800Y

6144 = 1800Y − 921.6Y

6144 = 878.4Y

∴ Y = $\dfrac{6144}{878.4}$

= 6.99

The value of Y is approximately = 7

21. 25cm^3 of 0.02M KOH neutralized 0.03g of a monobasic organic acid having the general formula $C_nH_{2n+1}COOH$. What is the molecular formula of the acid?
(C = 12, H = 1, O = 16)

Solution

Since the acid is monobasic one mole of it will react completely with one mole of KOH. Now, let us find the amount of KOH in 25cm^3 of 0.02m solution

1000cm^3 of the solution contains 0.02 mole of KOH

∴ 25cm^3 of the solution will contain x mole of KOH

Solving for x

X = $\dfrac{25 \times 0.02}{1000}$ = 0.0005 mole

The amount of KOH, which reacted, with 0.03g of the acid is 0.0005 mole.

Since, 0.0005 mole of KOH reacts with 0.03g of the acid
1 mole of KOH will react with xg of the acid.

Solving for x

$$X = \frac{1 \times 0.03}{0.0005} = 60g$$

Hence the molar mass (relative molecular mass) of the acid = 60g

The general formula of the acid
$C_nH_{2n+1}COOH$

Let us find n

The molar mass of the acid $C_nH_{2n+1}COOH$
= 12n + 2n+ 1 +12 + (16 x 2) + 1
= 14n + 46

Remember, the molar mass of the acid $C_nH_{2n+1}COOH = 60$
That is 14n + 46 = 60
14n = 60 - 46
14n = 14

$$\therefore n = \frac{14}{14} = 1$$

If n = 1

The molecular formula of the acid will be CH_3COOH

22. If $20cm^3$ of distilled water is added to $80cm^3$ of 0.50 $moldm^{-3}$ hydrochloric acid what will be the new concentration?

Solution

We use

$M_1V_1 = M_2V_2$

> M1 = Initial concentration
> M2 = Final concentration
> V1 = Initial volume
> V2 = Final volume

Where, M_1 = Initial concentration = $0.50 \, moldm^{-3}$
M_2 = Final concentration = ?
V_1 = Initial volume = $80 cm^3$
V_2 = Final volume = $80 + 20 = 100 cm^3$

$$M_2 = \frac{M_1V_1}{V_2} = \frac{0.50 \times 80}{100}$$

$$= 0.4 \, moldm^3$$

The new concentration = $0.4 \, moldm^{-5}$

23. $17.30 cm^3$ of a solution A containing $3.70 gdm^{-3}$ of an acid H_2X, neutralized $25 cm^3$ of a solution B containing $2.15g$ of sodium hydroxide per $250 cm^3$ of solution
Calculate.
(i) Concentration of solution B in $moldm^{-3}$
(ii) Concentration of solution A in $moldm^{-3}$
(iii) Molar mass of the acid H_2X

Solution

(i) To calculate the concentration of B in $moldm^{-3}$,

B is a solution containing 2.15g of sodium hydroxide per $250cm^3$.

Since, $250cm^3$ of solution B contains 2.15g of NaOH
$1000cm^3$ of solution B will contain Xg of NaOH.

We cross multiply, and solve for X

$$X = \frac{1000 \times 2.15}{250} = 8.6g$$

Hence, the concentration of B in gdm^{-3} = 8.6 gdm^{-3}

Remember

$$\text{Molar concentration} = \frac{\text{Concentration in } gdm^{-3}}{\text{Molar mass}}$$

Molar Concentration of B = $\frac{8.6}{50}$ = 0.215 $moldm^{-3}$

The concentration of B in $moldm^{-3}$ = 0.215 $moldm^{-3}$

(ii) We can calculate the concentration of A in $Moldm^{-3}$, using

$C_A V_A / C_B V_B = N_A / N_B$

The equation for the reaction

$H_2X\ (aq) + 2NaOH\ (aq) \rightarrow Na_2X\ (aq) + H_2O\ (l)$ from the equation

2 moles of NaOH reacts with 1 mole of H_2X

$C_A = C_B V_B N_A / V_A N_B$

$C_B = 0.215 \text{ moldm}^{-3}$,
$N_A = 1$
$V_A = 17.30 \text{ cm}^3$
$N_B = 2$
$V_B = 25 \text{ cm}^3$

$C_A = \dfrac{0.215 \times 25 \times 1}{17.30 \times 2} = 0.155 \text{ moldm}^{-3}$

(iii) To find the molar mass of the acid, H_2X

Remember,

Concentration in gdm^{-3} = Molar concentration x molar mass.

Hence, molar mass = $\dfrac{\text{Concentration in gdm}^{-3}}{\text{Molar concentration}}$

Concentration of H_2X in $\text{gdm}^{-3} = 3.70 \text{ gdm}^{-3}$

Molar concentration of $H_2X = 0.155 \text{ moldm}^{-3}$

Molar mass of $H_2X = \dfrac{3.70}{0.155}$

= 23.9g

The molar mass of H_2X is approximately = 24g

24. 25 cm^3 of an 0.09 moldm^{-3} solution of NaOH is neutralized by 22.6 cm^3 of an acid solution containing 4.5g of the acid per dm^3 of solution. Deduce the mole ratio of acid to alkali given that the molar mass of the acid is 90g.

Solution

First we determine the molar concentration of the acid, using

Molar concentration = $\dfrac{\text{Concentration in gdm}^{-3}}{\text{Molar mass}}$

= $\dfrac{4.5}{90}$ = 0.05 moldm^{-3}

Now using

$\dfrac{C_A V_A}{C_B V_B} = \dfrac{N_A}{N_B}$

Where N_A/N_B = mole ratio of acid to alkali

$C_A = 0.5 \text{moldm}^{-3}$ $V_A = 22.6 \text{cm}^3$

$C_B = 0.09 \text{moldm}^3$ $V_B = 25 \text{ cm}^3$

Substituting,

$\dfrac{0.05 \times 22.6}{0.09 \times 25} = \dfrac{\text{Mole of acid}}{\text{Mole of Base}}$

$\dfrac{\text{Mole of Acid}}{\text{Mole of Base}} = \dfrac{1.13}{2.25}$

Simplifying we have

$\dfrac{\text{Mole of Acid}}{\text{Mole of Base}} = \dfrac{1}{2}$

∴ Mole ratio of acid to base = 1:2

25. A piece of zinc was added to 1000cm^3 of 0.2M hydrochloric acid. After effervescence had stopped, 28cm^3 of the resulting solution required 17cm^3 of 0.08M sodium trioxocarbonate(iv) solution for complete neutralization.

Calculate the mass of the zinc added (Zn = 65, HCl = 36.5, Na_2CO_3 = 106)

$2HCl\ (aq) + Na_2CO_3\ (aq) \rightarrow 2NaCl\ (aq) + H_2O\ (l) + CO_2\ (g)$

Solution

First we need to find the actual concentration of HCl acid, which neutralized the Na_2CO_3. Using

$$\frac{C_A V_A}{C_B V_B} = \frac{N_A}{N_B}$$

C_A = ? C_B = 0.08M
V_A = 28cm³ V_B = 17cm³

$$C_A = \frac{C_B V_B N_A}{V_A N_B}$$

= (0.08 × 0.017 × 2)/(0.028 × 1) = 0.057 M

The actual concentration of HCl acid, which neutralized the Na_2CO_3
= 0.057 M.

If the resulting concentration of the solution of HCl after it had reacted with Zinc was 0.057M, it means the number of moles of the acid which reacted with zinc was 0.2M – 0.057M
= 0.143M.

The equation for the reaction between zinc and the acid is as follows

$2HCl\ (aq) + Zn\ (s) \rightarrow ZnCl_2\ (aq) + H_2\ (g)$

From this equation
2 moles of HCl is completely neutralized by 1 mole of Zn

That is 2 Moles of HCl is neutralized by the mass of 1 mole of Zinc (65g)

If 2 Moles of HCl is neutralized by the mass of 65 g of Zinc

Then, 0.143moles of HCl will be neutralized by x g of Zn

(we cross multiply)

X = (0.143 x 65)/2 = 4.648g

Therefore the mass of Zinc which was added to HCl = 4.648g

Questions

1. What is the relative molecular mass of a compound, which has empirical formula CH_2O?
[H = 1, C = 12, O = K]

2. $27cm^3$ of a $0.09 moldm^{-3}$ solution H_2SO_4 neutralized 25cm3 of solution containing 3.2g of impure NaOH in $250cm^3$ of solution. Calculate the percentage impurity of the NaOH sample?

3. 4.1g $Na_2CO_3.XH_2O$ crystals were made up to $250cm^3$ of solution with distilled water. $25cm^3$ of the resulting solution was neutralized by $27.5cm^3$ of a 0.1M solution of HCl. Determine the value of x?

SOLUBILITY

The solubility of a solute in a solvent is the maximum amount of the solute in moles or grams that will saturate I dm^3 of the solvent at a particular temperature.

A saturated solution of a solute is one, which has dissolved as much of the solute as it can dissolve at a given temperature in the presence of undissolved solute particles.
A supersaturated solution is one, which contains more of the solute than it can normally hold at that temperature.
A molar solution is one, which contains one mole of the solute a weight equivalent of the molar mass of the solute in one dm^3 of the solution.

Generally, solubility of a solute varies directly with temperature. This means the solubility of a solute increases with an increase in temperature and vice versa. So a solution would contain more of the solute at higher temperature, but when, cooled some of the solute would crystallize (precipitate) out of the solution.
Once we have understood and can apply these principles then we are well armed to face any mathematical problems on solubility.

Exercises

1. If 13g of magnesium trioxonitrate (v) was dissolved in 24g of distilled water at 30°c, calculate the solubility of the solute in $moldm^{-3}$.

Solution

First, we convert the mass of the solute to mole

Mole = $\dfrac{mass}{Molar\ mass}$

The molar mass of $Mg(NO_3)_2$ = 148

Mole = $\dfrac{13}{148}$ = 0.088 mole

> **Remember**: *Relative Density of Water = 1.*
> *This means that 1 g of water is equivalent to 1 cm^3 of water*

Therefore, 24g of water is equivalent to 24 cm^3 of water
And solubility is always in mole or gram per dm^{-3}

Since 24cm^3 of water contains 0.088 mole
1000cm^3 of water will contain x mole
we cross multiply and solve for x

x = $\dfrac{1000 \times 0.088}{24}$

∴ The solubility of $Mg(NO3)_2$ in water at 30°c is 3.66$moldm^{-3}$

2. 1.33dm³ of water at 70°C are saturated by 2.25 moles of lead (ii) trioxonitrate (v), $Pb(NO_3)_2$, and 1.33dm³ of water at 18°C are saturated by 0.53 mole of the same salt.
If 4.50 dm³ of the saturated solution is cooled from 70°C to 18°C, Calculate the amount of solute that will be deposited in (a) moles (b) grams
(Pb = 207, N = 14, O = 16)

Solution
At 70°C 1.33dm³ of the solution contains 2.25 moles
At 18°C 1.33dm³ of the solution contains 0.53 moles
Solute deposited on cooling from 70°C to 18°C
= 2.25 - 0.53 = 1.72 moles
On cooling from 70°C to 18°C,
1.33dm3 of the solution deposited 1.72 moles
∴ 4.5dm³ of the solution will deposit x moles, on cooling from 70°C to 18°C
We cross multiply and solve for x

x = $\underline{4.5 \times 1.72}$
　　　1.33　　= 5.82 moles

5.82 moles of the solute will be deposited.

We can convert mole to mass by applying
Mole = Mass/Molar mass
The molar mass of $Pb(NO_3)_2$ = 331,

Mass = mole x molar mass

Mass = 5.82 x 331 = 1926.4g

The mass of $Pb(NO_3)_2$ that will be deposited is 1926.4g or 1.93kg

3. Water was added to 120.0g of a salt MCl_2 to produce 50.0cm³ of a saturated solution at 25°C. Its solubility at 25°C is 8.0 moldm⁻³. Calculate the mass of the salt which remained undissolved. (M = 24, Cl = 35.5)

Solution

Let us first convert the solubility of MCl_2 to grams per dm³

(Molar mass of MCl_2 = 95)

Mass = mole x molar mass

 = 8.0 x 95   = 760g per dm³

$\boxed{1\ dm^3 = 1000cm^3}$

Since, 1000cm³ contains 760g of MCl_2

60cm³ will contain x g of MCl_2

X = $\dfrac{60 \times 760}{1000}$ = 45.6g

Only 45.6g of the salt dissolved to form 60cm³ of the solution.

Therefore the undissolved salt would be 120g – 45.6g = 74.4g

4. 90.0g of $MgCl_2$ was placed in $50.0cm^3$ of water to give a saturated solution at 298K. If the solubility of the salt is 8.0 $moldm^{-3}$ at the same temperature, what is the mass of the salt left undissolved at the given temperature?
(Mg = 24 , Cl = 35.5)

Solution
Let us convert the solubility of $MgCl_2$ to grams per dm^3
Applying, grams per dm^{-3} = $moldm^{-3}$ x molar mass
The molar mass of $MgCl_2$ = 95

Solubility of $Mgcl_2$ in grams per dm^3
= 8.0 x 95 = 760g per dm^3

Since $1000cm^3$ of the solution dissolved 760g of $MgCl_2$
$50cm^3$ of the solution will dissolve xg of $MgCl_2$
Cross-multiplying and solving for x
X = $\underline{50 \times 760}$
       1000   = 38g
Only 38g of $MgCl_2$ dissolved in $50cm^3$ of the solution

Therefore the mass of the undissolved salt will be
90.0 -38.0 = 52g

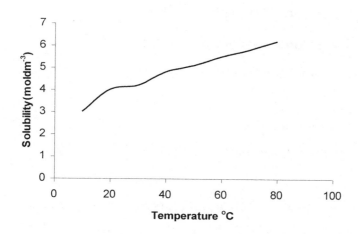

5. The diagram above is the solubility curve of a solute x. Find the amount of x deposited when 500cm³ of a solution of x is cooled from 60°C to 20 ° C

Solution
When faced with a problem like this one, trace the solubility of the salt at 60°c and 20°c from the curve
From the curve,
The solubility of x at 60°C is 5.5 moldm^{-3},
at 20°C it is 4.0 moldm^{-3}
That is to say,
At 60°C, 1000cm³ of the solution contains 5.5moles
And, at 20°c 1000cm³ of the solution contains 4.0moles.

Solute deposited on cooling from 60°c to 20°c
= 5.5 − 4.0 = 1.5moles

On cooling from 60°c to 20°c,

1000cm³ of the solution deposits 1.5 moles

∴ 500cm³ of the solution will deposit x moles

cross multiplying and solving for x

x = 500 x 1.5
    1000   = 0.75moles

The amount of solute deposited by 500cm³ is 0.75mole.

6. The solubility of sodium tetraoxosulphate (vi) is 1900g per 1000g of water at 80°C and 800g per 1000g water at 40°C. Calculate the mass of sodium tetraoxosulphate (vi) that will crystallize out of solution if 160g of the saturated solution at 80°C is cooled to 40°C.

Solution

The mass of the solution at 80°C = 1900g + 1000g = 2900g

The mass of the solution at 40°C = 800g + 1000g = 1800g

Solute deposited on cooling from 80°C to 40°C
= 2900 − 1800 = 1100g

On cooling from 80°C to 40°C

2900g of saturated solution deposits 1100g of solute

∴ 160g of saturated solution will deposit xg of solute

Cross – multiplying and solving for x

$$x = \frac{160 \times 1100}{2900} = 60.69g$$

On cooling from 80°C to 40°C 160g of saturated solution will deposit 60.69g of solute.

7. If the solubility of $KHCO_3$ is $0.40 Moldm^{-3}$ at room temperature, calculate the mass of $KHCO_3$ in $100cm^3$ of the solution at this temperature.
$[KHCO_3 = 100gmol^{-1}]$

Solution

The solubility of $KHCO_3$ = 0.40 $moldm^{-3}$

This means that,

1 dm ($1000cm^3$) of the solution contains 0.40 mole of $KHCO_3$

∴ $100cm^3$ of the solution will contain X mole of $KHCO_3$

Cross-multiply and solve for X

$$X = \frac{100 \times 0.40}{1000} = 0.04 \text{ mole of } KHCO_3$$

$100cm^3$ of the solution contains 0.04 mole of $KHCO_3$

Now let us convert the mole to mass. Remember,

Mass = Mole x Molar mass

Mass = 0.04 x 100 = 4g

Hence, $100cm^3$ of the solution contains 4g of $KHCO_3$.

8. Calculate the solubility of Na_2CO_3 at $25\,°C$, if $20.0cm^3$ of its saturated solution at that temperature gave 1.75g of the anhydrous salt.
[C = 12, O = 16, Na = 23]

Solution

Remember, that the solubility of a solute is usually measured in mole per dm^3 or gram per dm^3.
If $20cm^3$ of the saturated solution gave 1.75g of Na_2CO_3, it means it contained 1.75g per $20cm^3$.
If $20cm^3$ of saturated solution contains 1.75g of Na_2CO_3,
Then $1000cm^3$ of saturated solution will contain Xg of Na_2CO_3
Cross-multiply and solve for X
X = $\underline{1000 \times 1.75}$
          20   = 87.5g
The saturated solution contains 87.5g per dm^3. Converting to mol per dm^3, we apply,
$moldm^{-3}$ = gram dm^{-3}/Molar mass
molar mass of Na_2CO_3 = 106
$moldm^{-3}$ = 87.5/106 = 0.83 $moldm^{-3}$
The solubility of Na_2CO_3 at $25\,°C$ = 0.83 $moldm^{-3}$

GAS LAWS

When presented with questions on gas laws, we have to first consider the values we are given to determine which law would be most suitable for solving the problem. For instance, if we are given a question where the pressure is constant while the volume and temperature is changing, the most suitable law to apply would be the Charles law. And if we have one, which involves the number of moles of the gas, its volume, pressure and temperature, we would consider the ideal gas equation.

So as a whole, it is necessary to have a good knowledge of all the gas laws.

BOYLE'S LAW

Boyle's law states that the volume of a given mass of gas is inversely proportional to its pressure provided that the temperature remains constant.

The mathematical expression

$$P_1V_1 = P_2V_2$$

Where P_1 = Initial pressure

P_2 = Final pressure

V_1 = initial volume

V_2 = final volume

CHARLES'S LAW

Charles's law states that the volume of a given mass of gas is directly proportional to its temperature in Kelvin, provided that pressure remains constant.

$$\frac{V_1}{T_1} = \frac{V_2}{T_2}$$

Where, V_1 = Initial volume

V_2 = Final volume

T_1 = Initial temperature

T_2 = Final temperature

To convert temperature to Kelvin,

$0°C = 273k$

$K = °C + 273$

$°C = K - 273$

General Gas Equation

$$\frac{P_1V_1}{T_1} = \frac{P_2V_2}{T_2}$$

Where: P_1 = Initial pressure

P_2 = Final pressure

T_1 = Initial temperature

T_2 = Final temperature

V_1 = Initial volume

V_2 = Final volume

Dalton's Law of Partial Pressure

The law of partial pressures states that if there is a mixture of gases which do not react chemically together, then the total pressure exerted by the mixture is the sum of the partial pressure of the individual gases that make up the mixture.

$$P_{Total} = P_A + P_B + P_C + P_D$$

If a mixture of gases A, B, C, and D is collected over water.

$$P_{total} = P_{gas} + P_{water\ vapour}$$

That is the total pressure is the sum of the pressure of the gas and the saturated vapour pressure for water at that temperature.

Ideal Gas Equation

$PV = nRT$

Where: P = Pressure

V = Volume

T = Temperature

R = Molar gas constant

n = no. of moles of gas

Relative vapour density of a gas

= mass of a given volume of the gas or vapour
   Mass of an equal volume of hydrogen

Vapour density

= ½ x relative molecular mass of gas or vapour.

Graham's law of Diffusion

Graham's law of diffusion states that, at a constant temperature and pressure, the rate of diffusion of a gas is inversely proportional to the square root of its density.

$R_1/R_2 = \sqrt{\rho_2/\rho_1}$

Where R_1 and R_2 are the rates of diffusion of the different gases, and ρ_1 and ρ_2 are their densities respectively.

$R_1/R_2 = \sqrt{M_1/M_2}$

Where R_1 and R_2 are the rates of diffusion of the different gases, and M_1 and M_2 are their relative molecular masses, respectively.

Exercises

1. A given mass of Nitrogen is $0.12 dm^3$ at $60°C$ and $1.0 \times 10^5 NM^{-2}$. Find its pressure at the same temperature if its volume is changed to $0.24 \, dm^3$.

Solution

Looking at this problem temperature is constant while pressure and volume varies. Therefore, we apply Boyle's law

$P_1V_1 = P_2V_2$

$P_1 = 1.01 \times 10^5 NM^{-2}$.

$V_1 = 0.12 dm^3$

$P_2 = ?$

$V_2 = 0.24 \, dm^3$

$1.0 \times 10^5 NM^{-2} \times 0.12 = P_2 \times 0.24$

$P_2 = \dfrac{1.01 \times 10^5 NM^{-2} \times 0.12 dm^3}{0.24\ dm^3}$

$P_2 = 5.05 \times 10^4 NM^{-2}$

The new pressure is $5.05 \times 10^4 NM^{-2}$

2. A sample of nitrogen occupies volume of $1\ dm^3$ at 500k and $1.01 \times 10^5 NM^{-2}$. What will its volume be at $2.02 \times 10^5 NM^{-2}$ and 400k.

Solution

Here pressure, volume and temperature are all varying. So, we use the general gas equation.

$$\dfrac{P_1 V_1}{T_1} = \dfrac{P_2 V_2}{T_2}$$

$P_1 = 1.01 \times 10^5 NM^{-2}$

$V_1 = 1 dm^3$

$T_1 = 500k$

$P_2 = 2.02 \times 10^5 NM^{-2}$

$V_2 = x$

$T_2 = 400K$

Substituting the values

$V_2 = \dfrac{P_1 V_1 T_2}{P_2 T_1}$

$V_2 = \dfrac{1.01 \times 10^5 NM^{-2} \times 1 \times 400k}{2.02 \times 10^5 NM^{-2} \times 500k}$

$V_2 = 0.4 \, dm^3$

The new volume is $0.4 \, dm^3$

3. $130 cm^3$ of a gas at 20^0C exerts a pressure of $750 mmHg$. Calculate its pressure if its volume is increased to $150 cm^3$ at $35^0 C$.

Solution

The pressure, volume and temperature are all varying, so we have to use the general gas equation.

$\dfrac{P_1 V_1}{T_1} = \dfrac{P_2 V_2}{T_2}$

$P_1 = 750 mmHg$

$T_1 = 20 + 273 = 293k$

$V_1 = 130 cm^3$

$P_2 = ?$

$V_2 = 150 \, cm^3$

$T_2 = 35 + 273 = 308k$

Substituting we have,

$P_2 = \dfrac{P_1 V_1 T_2}{T_1 V_2}$

$= \dfrac{750 \times 130 \times 308}{150 \times 298}$

$P_2 = 683.3 mmHg$

The new pressure is $683.3 mmHg$

4. $30 cm^3$ of hydrogen was collected over water at 27^0C and 80mmHg. If the vapour pressure of water at the temperature of the experiment was 10mmHg, calculate the volume of the dry gas at 760mmHg and 7^0 C.

Solution

When a gas is collected over water.

The actual pressure exerted by the gas

= The pressure at which its collected - vapour pressure of water.

P_{gas} = 780 -10 = 770mmHg

Using. $\dfrac{P_1V_1}{T_1} = \dfrac{P_2V_2}{T_2}$

Where P_1 = Initial pressure = 770mmHg

V_1 = Initial volume = $30cm^3$
T_1 = Initial temperature = 273 + 27 = 300k
P_2 = Final pressure = 760mmHg
V_2 = Final volume = x
T_2 = Final temperature = 273 + 7 = 280k

$V_2 = \dfrac{P_1V_1 T_2}{P_2T_1}$

Substituting

$V_2 = \dfrac{770 \times 30 \times 280}{760 \times 300}$ = $\dfrac{64680000}{228000}$

= $28.4\ cm^3$

The volume of the gas at 760mmHg and 7^0 C = 28.4 cm^3

5. A given amount of gas occupies 10.0dm^3 at 4 atm and 273^0 C. The number of moles of the gas present is? [Molar volume of a gas at s.t.p = 22.4 dm^3]

Solution

The Ideal gas equation

$PV = nRT$

Where, P = pressure of gas in atm

V = volume of gas in dm^3

n = no. of moles of gas

R = gas constant

T = temperature in K

Since the value of R is not given, we have to first derive it, using the ideal gas equation at s.t.p when 1 mole of the gas occupies a volume of 22.4 dm^3,

P = 1atm, V= 22.4 dm^3, n = 1 mole, T= 273k

Substituting in PV = nRT

$R = \dfrac{PV}{nT} = \dfrac{1 \times 22.4}{1 \times 273}$

R = 0.082atm dm^3 mol^{-1} k^{-1}

The number of moles of the gas present

$n = \dfrac{PV}{RT}$ P = 4atm

V = 10.0 dm^3

$$T = 273 + 273 = 546k$$
$$R = 0.082 \text{ atm dm}^3 \text{ mol}^{-1} \text{ k}^{-1}$$

$$n = \frac{4 \times 10.0}{0.082 \times 546}$$

$$n = \frac{40}{44.77}$$

$$= 0.89 \text{ mole.}$$

6. If sulphur (iv) oxide and methane are released simultaneously at the opposite ends of a narrow tube, the rates of diffusion RSO_2 and RCH_4 will be in what ratio. [S = 32, O = 16, C = 12, H = 1]

Solution

Remember from Graham's law the rate of diffusion of a gas is inversely proportional to the square root of its relative molecular mass.

That is, for the gases,

$R\,CH_4/RSO_2 = \sqrt{MSO_2/MCH_4}$

Where, RCH_4 = Rate of diffusion of methane

RSO_2 = Rate of diffusion of sulphur (iv) oxide

MSO_2 = Relative molecular mass of SO_2

MCH_4 = Relative molecular mass of methane.

The relative molecular mass of methane, $CH_4 = [12 + (1 \times 4)]$
= 16

The relative molecular mass of $SO_2 = 32 + (16 \times 2) = 64$

Substituting,

$RCH_4/RSO_2 = \sqrt{64/16}$

$RCH_4/RSO_2 = \sqrt{64}/\sqrt{16}$

$RCH_4/RSO_2 = 8/4 = 2/1$

$RCH_4/RSO_2 = 2/1$

So the rate of diffusion RCH_4 and RSO_2 are in the ratio of 1:2

7. $56.0 Cm^3$ of a gas at standard temperature and pressure, s.t.p. weighed 0.11g. What is the vapour density of the gas? [molar volume of a gas at s.t.p. = 22.4 dm^3]

Solution

Remember that 1 mole of every gas occupies 22.4 dm^3 at standard temperature and pressure, s.t.p

That is at s.t.p. 22.4 dm^3 contains a mass equal to the relative molecular mass of the gas.

Since 0.056dm of the gas contains 0.11g at s.t.p.

22.4dm of the gas will contain Xg at s.t.p.

We cross multiply and solve for X

X = 22.4 x 0.11/ 0.056

X = 44g

∴ The relative molecular mass of the gas is 44g.

Remember,

Vapour density of a gas = ½ x relative molecular mass

∴ VD = ½ x 44

VD = 22

Vapour density of the gas is 22.

8. What is the temperature of a given mass of a gas initially at 0°C and 9 atm, if the pressure is reduced to 3 atm at constant volume?

Solution

Using, $P_1V_1/T_1 = P_2V_2/T_2$

Since the volume is constant, we have

$$P_1/T_1 = P_2/T_2$$

Where, P_1 = initial pressure = 9 atm

P_2 = Final pressure = 3 atm

T_1 = Initial temperature = 273 + 0.273K

T_2 = Final temperature = ?

$T_2 = P_2T_1/P_1$

Substituting, T_2 = 3 x 273/9 = 91

The final temperature = 91 K

9. The ratio of the initial to final pressure of a given mass of gas is 1:15. Calculate the final volume of the gas if the initial volume was 300cm^3 at the same temperature.

Solution

Using, $P_1V_1 = P_2V_2$

Is the same as $P_1/P_2 = V_2/V_1$

Substituting the values,

$1/1.5 = V_2/300$

$V_2 = 300 \times 1/1.5$

$V_2 = 200$

The final volume = $200 Cm^3$

10. The partial pressure of oxygen in a sample of air is 452mmHg and the total pressure is 780 mmHg. What is the mole fraction of oxygen?

Solution

Considering Dalton's law of partial pressure.

$$P_{Total} = P_A + P_B$$

The mole fraction of oxygen = 452/780 = 0.579

11. An oxide XO_2 has a vapour density of 32. What is the atomic mass of X?

Solution

Remember, Vapour density = ½ Relative molecular mass.
So, the relative molecular mass of $XO_2 = 2 \times 32 = 64$

If the relative molecular mass of $XO_2 = 64$ and knowing that the atomic mass of oxygen = 16, then we have the mathematical equation

$XO_2 = 64$

$X + (16 \times 2) = 64$

$X + 32 = 64$

$X = 64 - 32$

$\therefore X = 32$

The atomic mass of X = 32.

12. A gas X diffuses twice as fast as a gas Y under the same conditions. If the relative molecular mass of X is 28, calculate the relative molecular mass of Y.

Solution

Using,

$Ry/Rx = \sqrt{(Mx/My)}$

The ratio of the diffusion of X to Y is 2:1

Substituting in the above formula

$½ = \sqrt{(28/My)}$

$(½)^2 = 28/My$

$¼ = 28/My$

$My = 28 \times 4$

$My = 112$

The relative molecular mass of Y = 112

13. The volume occupied by 1.58g of a gas at s.t.p. is 500cm^3. What is the relative molecular mass of the gas? [M.V. of gases = 22.4 dm^3]

Solution

Remember, that 1 mole (molar mass) of any gas occupies 22.4dm^3 at s.t.p.
Hence, if 0.50dm^3 contains 1.58g of the gas at s.t.p.
Then, 22.4dm^3 will contain Xg at s.t.p. that is the molar mass of the gas.
Cross multiplying and solving for X,
X = 22.4 x 1.58/0.5
X = 70.78
The relative molecular mass of the gas = 70.78.

14. If 30 cm^3 of oxygen diffuses through a porous plug in 7 seconds, how long will it take 60cm^3 of chlorine to diffuse through the same plug?
[O = 16, Cl = 35.5]

Solution

Using, $RCl_2/RO_2 = \sqrt{MO_2/MCl_2}$
Where, MCl_2 = molar mass of chlorine
MO_2 = molar mass of oxygen
RCl_2 = rate of diffusion of Chlorine
RO_2 = rate of diffusion of oxygen
The rate of diffusion of oxygen = 30cm^3/7 secs = 4.3 cm^{3}S^{-1}

Substituting the values,

$RCl_2/4.3 = \sqrt{32/71}$

$RCl_2/4.3 = \sqrt{32}/\sqrt{71}$

$RCl_2 = 4.3 \times 5.7/ 8.47$

$RCl_2 = 2.91$

The rate of diffusion of chlorine = 2.91

And the volume of chlorine gas is 60 cm^3, we can find the time, using

Rate of diffusion = Volume/Time

Substituting,

2.91 = 60/t

t = 60/2.91

t = 20.6 secs

The time taken for 60cm^3 of chlorine to diffuse is 20.6 seconds.

15. 9.60g of a gas X occupies the same volume as 0.30g of hydrogen under the same conditions. Calculate the molar mass of X   [H = 1]

Solution

First we have to find the volume occupied by the gases.

1 mole of H_2 occupies 22.4dm^3,

That is, 2g of H_2 will occupies 22.4dm^3

∴ 0.3g of H_2 will occupy z dm^3

Cross-multiply and solve for z

z = 0.3 x 22.4/2

= 3.36dm^3

The gases occupied a volume of 3.36dm^3.

Now we know that,

9.60g of the gas X occupies a volume of 3.36dm^3,.

And as we know 22.4dm^3 will contain 1 mole of X.

Remember that, the molar mass is equal to the mass of 1 mole.

So, let the molar mass of X = y

9.60g of X occupies 3.36dm^3

∴ yg of X will occupy 22.4dm^3

Cross-multiply and solve for y

y = 9.60 x 22.4/3.36

y = 64

∴ The molar mass of X = 64 gmol^{-1}

16. Calculate the vapour density of a triatomic gas X if its relative atomic mass is 16.

Solution

Remember,

Vapour density = ½ relative molecular mass

Since X is triatomic, that is X$_3$

The relative molecular mass of X$_3$ = 3 x 16 = 48

∴ the vapour density of X = ½ x 48 = 24

17. Given that the mass of one molecule of the gas X is 2.19×10^{-22} g, determine the molar mass of X. [Avogadro's constant = 6.02×10^{23}].

Solution

Let the molar mass = y
From Avogadro's hypothesis we know that
 1 mole of a gas contains 6.02×10^{23} molecules,
 That is the molar volume of a gas contains 6.02×10^{23} molecules.
Since the mass of 1 molecule is 2.19×10^{-22}.
∴ the mass of 6.02×10^{23} molecules will be y
y = $6.02 \times 10^{23} \times 2.19 \times 10^{-22}$
    = 13.18×10^{1} = 131.8
the molar mass of gas X = 131.8 gmol^{-1}

18. A given volume of methane diffuses in 20 seconds. How long will it take the same volume of sulphur (iv) oxide to diffuse under the same conditions?
[CH_4 = 16, SO_2 = 64]

Solution

Since we are not given the volume of the gases, we have to find the ratio of the rate of diffusion of methane to sulphur (iv) oxide.

Using, $RCH_4/RSO_2 = \sqrt{MSO_2/MCH_4}$

Where, RCH_4 = Rate of diffusion of methane

RSO_2 = Rate of diffusion of sulphur (iv) oxide

MSO_2 = Relative molecular mass of SO_2

MCH_4 = Relative molecular mass of methane

Substituting,

$RCH_4/RSO_2 = \sqrt{64/16}$

$RCH_4/RSO_2 = \sqrt{64}/\sqrt{16}$

$RCH_4/RSO_2 = 8/4 = 2/1$

$RCH_4/RSO_2 = 2/1$

So the rate of diffusion RCH_4 and RSO_2 are in the ratio of 1:2

∴ if methane diffuses in 20 seconds, SO_2 will diffuse in 40 seconds.

19. Calculate the mass of bleaching powder that will produce 400cm^3 of chlorine at 25 °C and a pressure of $1.20 \times 10^5 Nm^{-2}$? [O = 16, Cl = 35.5, Ca = 40, molar volume of gases at s.t.p. = 22.4 dm^3, at s.t.p. standard pressure = $1.01 \times 10^5 Nm^{-2}$].

Bleaching powder reacts with dilute HCl according to the following equation,

$CaOCl_2$ (s) + 2HCl (aq) → $CaCl_2$ (aq) + H_2O (l) + Cl_2 (g)

Solution

First we have to find the volume which chlorine will occupy at s.t.p., since reactions are carried out at s.t.p.

Using $P_1V_1/T_1 = P_2V_2/T_2$

Where, $P_1 = 1.20 \times 10$ Nm

$P_2 = 1.01 \times 10$ Nm

$V_1 = 400$ cm

$V_2 = ?$

$T_1 = 25 + 273 = 298$ K

$T_2 = 273$ K

$V_2 = P_1V_1T_2/P_2T_1$

$= \dfrac{1.20 \times 10^5 \times 400 \times 273}{1.01 \times 10^5 \times 298}$

$= 131040/300.98 = 435.4$ cm^3

So we can deduce the mass of $CaOCl_2$ that would yield 435.4cm^3 of Cl_2 at s.t.p.

From the equation 1 mole of $CaOCl_2$ yields 1 mole of Cl_2 at s.t.p.

Molar mass of $CaOCl_2$ = 127

that is 127g of $CaOCl_2$ yields 22400cm^3 of Cl_2 at s.t.p.

∴ Xg of $CaOCl_2$ will yield 435.4cm^3 of Cl_2 at s.t.p.

Solving for X,

X = 127 x 435.4/22400

X = 2.468g

∴ 2.468g of $CaOCl_2$ will yield 400cm^3 of Cl_2 at 25 °C and 1.20×10^5 Nm^{-3}.

Questions

1. If 0.75 mole of cyclopropane and 0.66 mole of oxygen are mixed in a vessel with a total pressure of 0.7 atmosphere, what is the partial pressure of oxygen in the mixture?

2. A gas contained in a 3 dm^3 cylinder was allowed to expand into another cylinder of volume 4.5 dm^3. If the pressure in the second cylinder is 85.6Kpa, what was the original pressure of the gas?

3. The combustion of a hydrocarbon CH occurs according to the equation,
$C_3H_8 + 5O_2 \rightarrow 3CO_2 + 4H_2O$
Calculate the volume of oxygen required for the complete combustion of 20dm^3 of the gas at 30°C. The reaction occurs at atmospheric pressure.

4. 2.63g of a compound vapourized at 130°C and contained in a flask of volume 250 cm^3 had a pressure of 1.36 atm. What is the molecular weight of the compound?

5. The density of a gas is determined at 20°C and 1 atmosphere pressure to be 0.771gdm^{-3}. Calculate the molecular weight of the gas.

6. Calculate the pressure of 0.25 moles of ammonia gas confined to a volume of 5.21dm^3 at 28°C assuming the gas behaves ideally.

7. Argon gas was found to effuse completely from a container in 26 seconds. Another gas under the same conditions was found to effuse in 27.3 seconds. What is the molecular weight of this gas?

OXIDATION NUMBER

The oxidation number of an element indicates its oxidation state. When an element gains or loses electrons in a reaction, this would be reflected by a change in its oxidation number.

The oxidation number of an element in a particular molecule or ion is the electrical charge it appears to have as determined by a set of arbitrary rules. With these rules it is possible to calculate the oxidation numbers for the elements in the reactants and products of a chemical change.

RULES FOR DETERMINING OXIDATION

1. Positive ions have positive oxidation numbers (eg. Na^+ = +1), and Negative ions have negative oxidation numbers (eg. O^{2-} = -2)
2. The oxidation number of all element in the free state is zero, that is in their uncombined state is zero. For instance N_2, O_2, Na, Cu, Cl_2, their oxidation states is all zero.
3. The oxidation state of an ion consisting of more than one element, is equal to the algebraic sum of the oxidation numbers of all the elements in the ion.

   For example, for the SO_4^{2-} ion

   [Oxidation No of S] + 4x [Oxidation No of O]

   = +6 + 4 x -2

$$= +6 + (-8)$$
$$= -2$$

4. The algebraic sum of the oxidation numbers of all the elements in a compound is zero.

   For example in H_2SO_4

   2 x [OX. No of H] x [OX no of S] x 4 x [OX. No of O]

   $(2 \times +1) + 6 + 4x - 2$

   $2 + 6 + (-8)$

   $8 + (-8) = 0$

5. In most compounds, the oxidation number of oxygen is -2, except in peroxides where its oxidation number is -1. While for hydrogen it is +1, except in hydrides where its oxidation number is -1. So taking all these rules into consideration we could determine easily, the oxidation number of an element in a compound/molecule or the oxidation state of a radical (group of ions).

Exercises

1. Calculate the oxidation number of

(a) Mn in $KMnO4$ (b) S in H_2g

(c) I in KIO_3 (d) S in $Na_2S_2O_3$

Solution

(a) Bearing in mind that the sum of the oxidation numbers of the elements in a compound/molecule is equal to zero. And the oxidation number of K = +1, O = -2

Let, the oxidation number of Mn be Z

∴ For $KMnO_4$

[OX no of k] + [OX no. of Mn] + 4 x [o+ no. of O]

= +1 + Z + (4x -2) = O

1 + Z – 8 = 0

Z = 8 – 1

Z = +7

∴ The oxidation number of Mn in $KMnO_4$ is +7

(b) For H_2S

2x [OX no of H] + (OX. no of S] = 0

Let OX no of S be X

2 (+1) +X = 0

+2 + X = 0

∴ X = -2

∴ The oxidation number of S in H_2S is -2

(c) KIO_3 Let OX no of I = X

[OX no. of K] + [OX no of I] + 3x [OX no of O] = 0

(+1) + X + (3x -2) = 0

$1 + X + (-6) = 0$

$1 + X - 6 = 0$

$X = 6 - 1$

$X = +5$

∴ The oxidation number of I in KIO_3 is +5

(d) $Na_2S_2O_3$

Let the OX. No. of S be X

OX. No. of Na = +1

2x [OX No. of Na] + 2 x[OX No. of S] + 3x [OX No. of O] = 0

$2(+1) + 2X + 3(-2) = 0$

$2 + 2X - 6 = 0$

$2X = 6 - 2$

$2X = 4$

$X = 4/2 = 2$

$X = +2$

∴ The oxidation number of S in $Na_2S_2O_3$ is +2

2. Find the oxidation number of the chromium atom in potassium heptaoxodichromate (vi), $K_2Cr_2O_3$

Solution

Let Ox. No. of Cr be X

2x[Ox. No. of K] + 2x[Ox. No. of Cr] + 7x [Ox. No. of O]

$X = 0$

$(2x + 1) + 2X + (7x - 2) = 0$

$2 + 2 + - 14 = 0$

$2X = 14 - 2$

$2X = 12$

$X = \dfrac{12}{2}$

$= 6$

The oxidation number of Cr in $K_2Cr_2O_7$ is +6

3. Calculate the number of mole of electrons involved in the oxidation of 2.8g of iron fillings to iron (ii)ions. (Fe = 56).

Solution

First we write the equation of the oxidation of iron

$Fe (g) \rightarrow Fe^{2+} + 2e^-$

From the equation 2 moles of electrons are involved when 1 mole of Fe is oxidized to Fe^{2+}.

That is 2 mole of electrons are involved in the oxidation of 56g of Fe to Fe^{2+}

If 2 mole of electrons is involved in 56g of Fe

∴ x mole of electrons would be involved in 2.8g of Fe

Solve for x

$x = \dfrac{2 \times 2.8}{56} = 0.1$

In the oxidation of 2.8g of Fe to Fe^{2+} 0.1 mole of electron is involved.

4. What is the oxidation number of Nitrogen in $Al(NO_3)_3$?

Solution

We know that Oxidation number of Al = +3, O = -2

Let the oxidation number of nitrogen be n

+3 + 3(n + 3x -2) = 0

3 + 3(n – 6) = 0

3 + 3n – 18 = 0

3 + 3n = 18

3n = 18 – 3

3n = 15

∴ n = 15/3 = +5

The oxidation state of nitrogen in $Al(NO_3)_3$ = +5

ELECTROLYSIS

Electrolysis is the chemical decomposition of a compound resulting from the passage of electric current through a solution or molten form of the compound.

Faraday's first law of electrolysis

Faraday's first law of electrolysis states that the mass of an element discharged during electrolysis is directly proportional to the quantity of electricity passing through it.

The quantity of electricity, Q is measured in coulombs

Q = I x t   I = Current (in amperes)

                t = time (in seconds)

Faraday's 1st law

Mass, M = EIt

E = Constant (electrochemical equivalence)

Calculations on electrolysis do not require a complete knowledge of formulas, but an understanding of the principles like, the Faraday is the minimum quantity of electricity required to liberate one mole of singly charged ions (univalent elements) and it has a value of 96500C.

The discharge of a mole of a univalent element involves the transfer of a mole of electron. Therefore one faraday is equivalent to a mole of electron.

Faraday's second law

It states that when the same quantity of electricity is passed through different electrolytes, the relative number of moles of the elements discharged are inversely proportional to the charges on the ions of the elements.

Exercises

1. Calculate the mass of silver deposited when a current of 3.0A flows through a solution of a silver salt for 80 minutes.
(Ag = 108, 1 Faraday = 965000C)

Solution

First, let us calculate the quantity of electricity used.
Q = I x t = 3.0 x 80 x 60 = 14400C
The equation for the discharge of silver is
$Ag^+(aq) + e^- \rightarrow Ag(s)$

1F discharges 108g

We can say that one Faraday discharges 1 mole of silver that is 108g of silver.
Since, IF (96500C) discharges 108g of silver
144800C will discharge xg of Ag
We cross multiply and solve for X
X = 14400 x 108
        96500   =   16.1g

The mass of silver deposited is 16.1g

2. 1.2F of electricity are passed through electrolytic cells containing Na^+, Cu^{2+}, and Al^{3+} in series. How many moles of each metal would be formed at the cathode of each cell?

Solution

The equation for the discharge of Na^+

$$Na^+(aq) + e^- \rightarrow Na(s)$$

From the equation one mole of sodium requires a faraday to be discharged

Since 1F discharges 1 mole of Na

1.2F will discharges x mole of Na

cross multiplying and solving for x

x = $\dfrac{1.2 \times 1}{1}$ = 1.2

1.2 mole of Na

The equation for the discharge of Cu^{2+}

$$Cu^{2+}(aq) + 2e^- \rightarrow Cu(s)$$

From the equation 1 mole of Cu requires 2 faradays to be discharged.

Since 2 faradays discharges 1 mole of Cu

1.2 F will discharge x mole of Cu

Cross multiplying and solving for x

x = $\dfrac{1.2 \times 1}{2}$

= 0.6

0.6 mole of Cu

The equation for the discharge of Al^{3+}

$Al^{3+}(aq) + 3e^- \rightarrow Al(s)$

From the equation 1 mole of Al requires 3 faradays to be discharged.

Since 3F discharges 1 mole of Al

1.2F will discharge x mole of Al

$x = \dfrac{1.2 \times 1}{3}$ = 0.4 mole

0.4 mole of Aluminium

3. What mass of Gold is deposited during the electrolysis of gold (iii) tetraoxosulphate (vi) when a current of 15A is passed for 193 seconds?

(Au = 97, F = 96500C).

Solution

The quantity of electricity used,

Q = I x t

Q = 15 x 193 = 2895C

The equation for the discharge of gold is

$Au^{3+}(aq) + 3e^- \rightarrow Au(s)$

From the equation 3 faradays are required to discharge one mole of gold, i.e. 97g of Au

Since 3F (3 x 96500C) discharges 97g of Au
2895C will discharge Xg of Au
We cross multiply and solve for x

$$X = \frac{2895 \times 97}{3 \times 96500} = \frac{280815}{289500}$$

$$= 0.97g$$

The quantity of gold deposited is 0.97g

4. What current in amperes will deposit 2.7g of aluminium in 2 hours?

(AL = 27, F = 96500C)

Solution

When faced with a question like this, we should first determine the quantity of electricity, which would deposit 2.7g of Al. The equation for the discharge of Al is $Al^{3+}(aq) + 3e^- \rightarrow Al(s)$

From the equation 1 mole of Al is deposited by 3 faradays of electricity.

Since 3 faradays deposits 1 mole of Al
i.e. 3 x 96500C deposits 27g of Al
Then x C of electricity will deposit 2.7g of Al
Cross-multiplying and making x the subject

$$X = \frac{3 \times 96500 \times 2.7}{27}$$

= 28950C

The quantity of electricity that deposited 2.7g of Al was 28950C.

This means that, the quantity of electricity, which flowed in 2 hours was 28950C. So we can now calculate the current.

Remember,
Q = I x t   t = 2 x 60 x 60 secs
Substituting
28950 = I x 2 x 60 x 60
∴ I = $\dfrac{28950}{2 \times 60 \times 60}$
　　= 4.02A

The current was 4.02 A

5. What amount of mecury would be liberated if the same quantity of electricity that liberates 0.65g of zinc is applied?
(Zn = 65, Ag = 201, F = 96500C)

Solution

First we have to find the quantity of electricity, which deposited 0.65g of zinc.

The equation for the discharge of zinc
$Zn^{2+}(aq) + 2e^- \rightarrow Zn(s)$

From the equation 1 mole of zinc requires two faradays to be discharged. Since 2 faradays discharges 1 mole of zinc i.e.
2 x 96500C discharges 65g of zinc

XC will discharge 0.65g of zinc

$$X = \frac{2 \times 96500 \times 0.65}{65}$$

= 1930C

The quantity of electricity, which liberated 65g of zinc is 1930C.

For mercury

$Hg^+(aq) + e^- \rightarrow Hg(s)$

1 mole of mercury requires 1 Faraday to be liberated.
Since 1 faraday liberates 1 mole (201g) of Hg
i.e. 96500C liberates 201g of Hg
Then 1930C will liberate Xg of Hg
Cross multiplying and solving for x

$$X = \frac{1930 \times 201}{96500}$$

= 4.02g of mercury

6. In the reaction below, calculate the quantity of electricity required to discharge zinc.

½ Zn^{2+}(aq) + e^- → ½ Zn(s)

Solution

Remember, A faraday is the quantity of electricity, which is required to discharge one mole of a univalent element.

Normally, two faradays would be required to liberate one mole of zinc since zincs is a divalent element. So to discharge half a mole of zinc in the above equation, only one faraday is required, that is 96500C.

7. How many faradays of electricity are required to deposit 0.20 mole of Nickel, if 0.10 faraday of electricity deposited 2.98g of nickel during electrolysis of its aqueous solution?

[N_1 = 58.7, 1F = 96500C]

0.10 faraday deposits 2.98g of Nickel changing this mass to mole, by using

$$\text{Mole} = \frac{\text{mass}}{\text{Molar mass}} = \frac{2.98}{58.7}$$

$$= 0.051 \text{ mole}$$

So, 0.10 F deposits 0.051 mole of Nickel
X F of electricity will deposit 0.20 mole of Nickel
If we cross-multiply and solve for x

$$X = \frac{0.10 \times 0.20}{0.051} = 0.39$$

∴ 0.39 faradays of electricity will deposit 0.20 mole of Nickel

8. 0.22g of a divalent metal is deposited when a current of 0.45A is passed through a solution of its salt for 25 minutes. Calculate the relative atomic mass of the metal.

Solution

First let us calculate the quantity of electricity that flowed.

Q = It
= 0.45 x 25 x 60
= 675 coulombs

Since the metal is divalent, we could represent its discharge as follows

$M^{2+}(aq) + 2e^- \rightarrow M(s)$

That means 2 faradays will discharge 1 mole of the metal

The mass of one mole of an element is equivalent to the relative atomic mass of the element.

675 coulombs deposits 0.22g of the metal

∴ 2 x 96500C will deposit Xg (i.e. the relative atomic mass of the metal)

$X = \dfrac{2 \times 96500 \times 0.222}{675} = \dfrac{42846}{675}$

= 63.5g

The relative atomic mass of the divalent metal is 63.5g.

9. A current is passed through three electrolytic cells connected in series, containing solutions of silver trioxonitrate (v), copper (ii) tetraoxosulphate (vi) and brine respectively. If 12.7g of copper are deposited in the second electrolytic cell, calculate.

    (a) the mass of silver deposited in the first cell,

    (b) the volume of chlorine liberated in the third cell at 17°C and
        800mmHg pressure

(Ag = 108, Cu = 63.5, 1F = 96500C, G.M.V. of gases of S.T.P. = 22.4dm^3)

Solution

When faced with a problem like this, we first find the quantity of electricity, which flowed. This we can find by considering the discharge of copper.

$Cu^{2+}(aq) + 2e^- \rightarrow Cu(s)$

From the equation

2 Faradays of electricity discharge 1 mole copper

That is 2 x 96500C discharge 63.5g of Cu

∴ X Coulombs discharged 12.7g of Cu

We cross multiply and solve for X

X = $\frac{2 \times 96500 \times 12.7}{63.5}$ = $\frac{2451100}{63.5}$

=38600

The quantity of electricity that flowed was 38600C

(a)   For silver

   $Ag^+(q) + 2e^- \rightarrow Ag(s)$

   From the equation

1 faraday discharges 1 mole of silver

That is 96500C discharges 108g of silver

∴ 38600C will discharge xg of silver

(since the same quantity of electricity flows through the three cells).

We cross multiply and solve for X

$$X = \frac{38600 \times 108}{96500} = \frac{4168800}{96500} = 43.2$$

The mass of silver deposited was 43.2g

(b) The equation for the discharge of chlorine is

$Cl^- \rightarrow Cl + e^-$

$Cl + Cl \rightarrow Cl_2$

This shows that for the discharge of 1 mole of chlorine gas two faradays are involved. Remember, that at s.t.p one mole of chlorine gas occupies 22.4dm^3 so we can say that

2 x 96500C liberates 22.4 dm^3 of Cl_2 of s.t.p

∴ 38600C will liberate X dm^3 of Cl_2 at s.t.p.

$$X = \frac{38600 \times 22.4}{2 \times 96500} = \frac{864840}{193000} = 4.48 dm^3$$

4.48dm^3 of chlorine gas was liberated at s.t.p.

Using, the general gas equation,

$$\frac{P_1V_1}{T_1} = \frac{P_2V_2}{T_2}$$

We can calculate the volume of chlorine liberated at 17°C and 800mmHg

P_1 = 760mm Hg (Standard pressure)

T_1 = 273k (standard temperature)

V_1 = 4.48dm^3

P_2 = 800mmHg
T_2 = 273 + 17 = 290k
V_2 = ?

$V_2 = \dfrac{T_2 P_1 V_1}{P_2 T_1}$

$= \dfrac{760 \times 4.48 \times 290}{800 \times 273}$

$= \dfrac{987392}{218400} = 4.52 dm^3$

The volume of chlorine liberated at $17^{\circ}C$ and 800mmHg is 4.52 dm^3.

10. A current of 0.72 amperes was passed through dilute H_2SO_4 acid for 3 hours 20 minutes. Calculate
(i) The quantity of electricity that was passed
(ii) If 1 dm3 of gas was evolved at the cathode during the electrolysis of acidified water, what was the volume of the gas evolved at the anode?

Solution

Quantity of electricity = Current x time (in seconds)
Q = 0.72 x (3 x 60 x 60) + (20 x 60)
Q = 0.72 x 12,000g
Q = 8640 coulombs

The quantity of electricity that was passed = 8640 coulombs

(ii) To answer this question we have to first, write the cathodic and anodic half reactions for the electrolysis of acidified water.

Cathodic Half reaction

$4H^+(aq) + 4e^- \rightarrow 2H_2(g)$

Anodic Half reaction

$4OH^-(aq) \rightarrow 2H_2O(l) + O_2(g) + 4e^-$

From the equations above we can see that when 2 moles of Hydrogen is discharged at the cathode, 1 mole of oxygen is also discharged at the anode.

Therefore, when $1dm^3$ of Hydrogen is discharged at the cathode, $0.5dm^3$ of oxygen will also be discharged at the anode.

12. A current of 0.72 amperes was passed through dilute tetraoxosulphate (vi) acid for 3 hours 20 minutes, calculate the quantity of electricity that was passed. If $1dm^3$ of gas was evolved at the cathode during electrolysis of acidified water, what was the volume of gas evolved at the anode?

Solution

The quantity of electricity can be calculated using,

$\qquad Q \quad = \quad It$

$\qquad t \quad = \quad$ 3 hours 20 mins.

$\qquad \quad \quad = \quad$ 3 x 60 x 60 x 20 x 60

$$= 10800 \times 1200 = 120{,}000 \text{ seconds}$$
$$Q = 0.72 \times 120{,}000 = 8640 C$$

To answer this question, one must know very well, the electrolysis of acidified water.

During the electrolysis of acidified water for every volume of oxygen liberated at the anode, two volumes of hydrogen are liberated at the cathode.

Therefore, when 1 dm^3 of gas is evolved at the cathode, half the volume, that is 0.5dm^3 of gas will be evolved at the anode.

13. Determine how many moles of electrons are transferred when 4825 coulombs of electricity is passed through an electrolytic cell?

[1F = 96500C)

Solution

Remember, that when 1 faraday that is 96500C of electricity flows, 1 mole of electron is transferred.

Since, the flow 96500C of electricity involves transfer of 1 mole of electron then, the flow of 4825C of electricity will involve the transfer of X mole of electrons.

Cross-multiply and solve for X

$$X = \frac{4825 \times 1}{96500} = 0.05$$

0.05 mole of electron is transferred when 4825 coulombs of electricity flows.

14. Calculate the number of copper (ii) ions that will be discharged by 0.250F.
[Avogadro constant = 6.02×10^{23} mol^{-1}]

Solution

The equation for the discharge of copper (ii) ions
$Cu^{2+}(aq) + 2e^- \rightarrow Cu(s)$

From the equation 2 faradays of electricity discharges 1 mole of copper (ii) ions
Remember according to Avogadro,
1 mole of copper (ii) ion contains 6.02×10^{23} copper (ii) ions
So, 2 faradays discharges 6.02×10^{23} copper (ii) ions
∴ 0.250 faradays will discharge x copper (ii) ions
Cross multiply and solve for X
X = $\underline{0.250 \times 6.02 \times 10^{23}}$
              2
= 7.5×10^{22} ions
0.250 F of electricity discharges 7.5×10^{22} copper (ii) ions

15. A Voltmeter Containing silver trioxonitrate (v) solution was connected in series to another voltmeter containing copper (ii) tetraoxosulphate (vi) solution. When a current of 0.200 ampere was passed through the solutions, 0.780g of silver was deposited, calculate the

(i) Mass of copper that would be deposited in the copper voltmeter

(ii) Quantity of electricity used and the time of current flow.

[Cu = 63.5, Ag = 108, 1F = 96500C]

Solution

First, we have to find the quantity of electricity that flowed, depositing 0.78g of silver.

Silver is univalent

$$Ag^+(aq) + e^- \rightarrow Ag(s)$$

From the equation of discharge of silver 1 mole of Ag requires 1F of electricity to be deposited.

That is 96500C of electricity deposits 108g of Ag

∴ X C of electricity deposited 0.78g of Ag

Cross multiply and solve for X

X = $\dfrac{96500C \times 0.78}{108}$ = 696.9C

The quantity of electricity that flowed is 696.9C

Using, Q = I x t, we can calculate the time taken

Q = 646.9C
I = 0.200 A
696.9 = 0.200 × t
t = $\frac{696.9}{0.200}$ = 3484 seconds

The time taken for the current to flow is 3483 secs (58 mins)

(i) The equation for the discharge of copper is
$Cu^{2+} + 2e^- \rightarrow Cu(s)$
Copper is divalent, from the equation 2 faradays of electricity is required to discharge one mole of copper.

Hence 2 × 96500C deposits 63.5g of copper
∴ 696.9C will deposit Xg of copper
Cross multiply and solve for X
X = $\frac{696.9 \times 63.5}{2 \times 96500}$ = 0.229g

The mass of copper deposited = 0.229g

16. 0.45g of a metal M was deposited when a current of 1.8 amperes was passed for 12.5 minutes through a solution containing M^{2+}. Calculate the relative atomic mass of M. [1 faraday = 96500C].

Solution

First, let us find the quantity of electricity that flowed.

Using, Q = I x t   t = 12.5 x 60 x 60 secs

Q = 1.8 x 12.5 x 60 = 1350C

Now we know that 1350C of electricity deposited 0.45g of M.

Since M is a divalent metal

$M^{2+}(aq) + 2e \rightarrow M(s)$

Therefore, 2 faradays of electricity will deposit 1 mole of M, hence a mass of M, which is equivalent to the molar mass of M. Remember, the molar mass of an element is equivalent to its relative atomic mass.

1350C of electricity deposits 0.45g of M

∴ 2 x 96500C of electricity will deposit Xg of M

Cross multiply and solve for X

$$X = \frac{2 \times 96500 \times 0.45}{1350} = 64.3g$$

Therefore, the relative atomic mass of M is 64.3g.

17. What mass of copper would be deposited by a current of 1.0 ampere passing for 965 seconds through copper (ii) tetraoxosulphate (vi) solution? [Cu = s63.5, 1F = 9600C]

Solution

Fist, we find the quantity of electricity that flowed,

Using Q = I x t
Q = 1.0 x 965 = 965C

The equation for the discharge of copper is
Cu^{2+} (aq) + 2e → Cu(s)
Copper is a divalent metal. From the equation 2 faradays will discharge 1 mole of copper.
That is 2 x 96500C of electricity deposits 63.5g of Cu
∴ 965 C of electricity will deposit Xg of Cu
Cross multiply and solve for X

X = $\dfrac{965 \times 63.5}{2 \times 96500}$ = 0.32g

The mass of copper deposited = 0.32g

ENERGY AND CHEMICAL REACTIONS

Energy (heat) changes during a reaction, is represented by the enthalpy change, ΔH.

Enthalpy change, ΔH

= Heat content of products, Hp – Heat content of reactants, Hr

$$\Delta H = Hp - Hr$$

The unit for measuring energy is the joule (J). One Joule is the energy required to raise the temperature of 1g of water by 0.239K (or ^0C).

If the Hp is greater than the Hr then ΔH will be positive, this means that heat will be absorbed from the environment. This gives rise to an endothermic reaction. If Hp is less than Hr, then ΔH will be negative and heat is released into the environment, this gives rise to an exothermic reaction.

Hess's Law and the Born-Haber Cycle

Hess's law of constant heat summation states that the total enthalpy change of a chemical reaction is constant regardless of the route by which the chemical change occurs, provided that the conditions at the start of a reaction are the same as the final conditions.

Using this law we can determine the enthalpy change of a chemical reaction theoretically from known values of enthalpy changes of related reactions.

For example, for the reaction we can use an energy cycle to represent Hess's law.

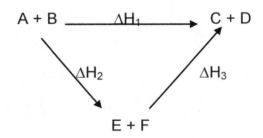

Here,

Heat change when = Sum of heat changes when
A + B → C + D   (A + B → E + F) and (E + F → C + D)

This law can also be represented by a Born-Haber cycle, which combines an energy cycle and energy level diagram.

Born – Haber Cycle

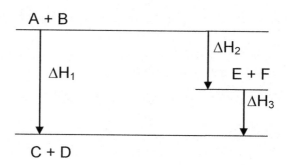

$\Delta H_1 = \Delta H_2 + \Delta H_3$

Exercises

1. In the formation of ethanol, C_2H_5OH

$2C\ (s) + 3H_2\ (g) + ½O_2\ (g) \rightarrow C_2H_5OH\ (l)$

This is a reaction that cannot be carried out experimentally. The enthalpy changes for the following reactions are however known.

a. $C_2H_5OH\ (l) + 3O_2\ (g) \rightarrow 2CO_2\ (g) + 3H_2O\ (l)$, $\Delta H = -1367$ KJ mol^{-1}

b. $C\ (s) + O_2\ (g) \rightarrow CO_2\ (g)$, $\Delta H = -1393.5$ KJ mol^{-1}

c. $H_2\ (g) + ½O_2\ (g) \rightarrow H_2O\ (l)$, $\Delta H = -285.9$ KJ mol^{-1}

Use the data to determine the enthalpy change in the formation of ethanol.

Solution

The first step is to determine a way of combining the reactions of known enthalpy changes, to produce ethanol by the given reaction.

We make the following observations,

(i) The equation has C_2H_5OH (l) on the right side, while equation a. has C_2H_5OH (l) on the left side. We put C_2H_5OH (l) on the same side and this can be done by reversing equation a.

(ii) The equation has C (s), H_2 (g) and O_2 (g) on the left side, so does equation b and c.

(iii) The equation has 2 moles of CO_2 while equation (b) has only 1 mole CO_2, we must make these equal by multiplying equation (b) by 2.

(iv) The equation has 3 moles of H_2O, while equation (c) has only 1 mole of H_2O, we make these equal by multiplying equation (c) by 3.

Following these we rewrite the equations

(a): $2CO_2$ (g) + $3H_2O$ (l) → C_2H_5OH (l) + $3O_2$ (g), $\Delta H = +1367$

2 x (b): $2C(s) + 2O_2(g)$ → $2CO_2$ (g), $\Delta H = 2 \times -393.5$ KJ mol^{-1}

3 x (c): $3H_2(g) + 3/2O_2(g)$ → $3H_2O(l)$, $\Delta H = 3 \times -285.9$ KJ mol^{-1}

Summing the three equations we have

$2C(s) + 7/2O_2(g) + 3H_2(g)$ → C_2H_5OH (l) + $3O_2(g)$

$\Delta H = (+1367 - 787 - 857.7)$ KJ mol^{-1}

Canceling out species that occur on both sides, we have

$$2C(s) + 3H_2(g) + \tfrac{1}{2}O_2(g) \rightarrow C_2H_5OH(l), \Delta H = -277.7 KJ-$$

BOND ENERGY

This is the average amount of energy associated with making or breaking 1 mole of a particular bond, in its gaseous state. We can use bond energies to calculate the heat/enthalpy change of a reaction.

For instance,

Calculate the approximate heat of reaction of the following reaction from the bond values given

$H_2(g) + Cl_2(g) \rightarrow 2HCl(g)$

H – H   =   435 KJ mol^{-1}

Cl – Cl =   244 KJ mol^{-1}

H – Cl   =   431 KJ mol^{-1}

Solution

Energy Change (KJ)

1 Mole of H –H bond broken   ½ x (+435)   =   217.5

1Mole of Cl – Cl bond broken   ½ x (+244)   =   112

2 Moles of H – Cl bond formed 1 x (-431)   =   - 431

217.5 + 112 = 329.5

$\Delta H_{reaction}$ = Total Energy required + total energy evolved

ΔHr = 329.5 − 431

ΔHr = −101.5KJ

CHEMICAL THERMODYNAMICS

For a system:

Change in internal = Heat absorbed by the system, q + work done by energy ΔU the system, W

q is positive when the system absorbs heat (energy) from the environment, and negative when it evolves heat into the environment. W is positive when the system absorbs energy by having work done on it by the surrounding and Negative when the system evolves energy by doing work on the surrounding.

That is

$\Delta U = q + W$

When the system is gaseous work done is given by $P\Delta V$

Where P = pressure

ΔV = Change in volume

$\Delta H = \Delta U + P\Delta V$

Given that the heat absorbed by the system at constant pressure is equal to the enthalpy change.

Entropy Change, ΔS

$$\Delta S = \frac{\Delta H}{T} \quad \text{Where } T = \text{Temperature}$$

Free energy change, ΔG

$$\Delta G = \Delta H = T\Delta S$$

Standard free energy, ΔG^θ

$$\Delta G^\theta = -RT\ln k$$

Where R = Gas constant
T = Temperature of system in Kelvin
K = Equilibrium constant

Standard electrode potential of a redox reaction in an electrochemical cell, E^θ

$$E^\theta = \frac{RT\ln k}{nF}$$

Where R = Gas constant
T = Temperature in Kelvin
K = Equilibrium constant
n = Number of moles of electrons
F = Faraday's constant

$$\Delta G^\theta = -nFE^\theta$$

Where ΔG^θ = Standard free energy change
n = Number of moles of electrons

F = Faraday's constant
E^θ = Standard Electrode potential

When ΔG^θ is negative, it means that work is obtained from the electrochemical cell.

Heat evolved during a reaction is determined by
Heat evolved = Mass x specific heat capacity x temperature rise
Heat change = Mass and specific heat capacity x temperature change.

RATE OF REACTIONS

Rate of reaction = $\dfrac{\text{Mass of reactant}}{\text{Time taken for the change}}$

Or

Rate of reaction = $\dfrac{\text{Volume of gaseous product formed}}{\text{Time taken for the change}}$

Average rate of whole reaction = $\dfrac{\text{Total loss of mass}}{\text{Total reaction time}}$

Rate of reaction = $\dfrac{\text{Change in concentration of reactant/product}}{\text{Time taken for the change}}$

Exercises

1. Find the rate of reaction, if 0.9g of calcium trioxocarbonate (iv) reacted with excess dilute hydrochloric acid to evolve carbon (iv) oxide, the entire reaction took 9 minutes.

Solution

Rate of reaction = $\dfrac{\text{Mass of reactant}}{\text{Time taken}}$

= $\dfrac{0.9g}{9mm}$ = 0.1

Rate of reaction = 0.1g min^{-1}

2. Calculate the change in internal energy for a gas that releases 35J of heat and has 128J of work done on it.

Solution

Remember

$\Delta U = q + w$ $q = -35J$

$\Delta U = -35 + 128$ $w = +128J$

$\Delta U = 93KJ$

3. Calculate the entropy change for the conversion of one mole of liquid water to vapour at 100°C, if the heat of vapourization of water = 2260.87 Jg^{-1}

Solution

The mass of 1 mole of water = 18g

Hence heat of vapourization of

1 mole of water = 2260.87 × 18 = 40695.70J

applying $\Delta S = \dfrac{\Delta H}{T}$ T= (100 + 273) = 373

$\Delta S = \dfrac{40693.70}{373}$

$= 109.10 JK^{-1} mol^{-1}$

Using the values of bond energies given, calculate the heat of reaction for the combination of acetylene according to the equation

166

$C_2H_2(g) + 5/2 O_2(g) \rightarrow 2CO_2(g) + H_2O(g)$.

$C \equiv C$,   837 KJ mol^{-1} $C = O$, 736 mol^{-1}
$H - C$,   414 KJ mol^{-1}
$O - O$,   498 KJ mol^{-1} $H - O$, 464 KJ mol^{-1}

Solution

The reaction could be represented as

$H - C \equiv C - H + 5/2 (O - O) \rightarrow 2(O=C=O) + H - O - H$

In this reaction 1 mol of $C = C$, 2 mol $H - C$ and 5/2 mol $O - O$ bonds are broken and 4 mol $C - O$ bonds and 2 mol $H - O$ bonds are formed.

Using the values.

Bond Breaking

1 Mol $C \equiv C$	=	1 × 837 KJ =	837
2 Mol $H - C$	=	2 × 414 KJ =	828
5/2 Mol O-O	=	5/2 × 498 KJ =	1245
		Total =	2910

Bond formation

4 mol $C = O$	=	4 × -736 =	-2944
2 mol $H - O$	=	2 × -464 =	-928
Total	=		-3872

Remember

$\Delta R_{eaction}$ = Summation of ΔH + Summation of ΔH
             of bonds broken   of bonds formed.

ΔH = 2910 – 3872
ΔH = -962KJ

4. 0.46g of ethanol when burned raised the temperature of 50g of water by 14.3k. Calculate the heat of combustion of ethanol. [C = 12, O = 16, H = 1, specific heat capacity of water = 4.2 Jg^{-1} K^{-1}].

Solution
Heat evolved by combustion of 0.46g of ethanol is equal to the heat absorbed by 50g of water.
Heat evolved = mass x specific heat capacity x rise in temperature
= 50g x 4.2Jg^{-1} x 14.3k
= 3003J

The mass of 1 mole of ethanol = 46g
The heat of combustion of ethanol is the amount of heat energy evolved by combustion of 1 mole of ethanol. Since the combustion of 0.46g of ethanol yields 3003 J.

∴ 46g of ethanol would yield XJ
Cross multiply and solve for X
X = $\dfrac{46 \times 3003}{0.46}$

$$= 300,300 \text{ J mol}^{-1}$$

∴ The heat of combustion of ethanol is
$$= -300.3 \text{ KJ mol}^{-1}$$

It carries a negative sign because combustion is an exothermic reaction.

5. The combustion of ethane, C_2H_4, is given by the equation
$$C_2H_4 + 3O_2 \rightarrow 2CO_2 + 2H_2O : \Delta H = -14.28 \text{ KJ}.$$
If the molar heats of formation of water and carbon (iv) oxide are -286 KJ and -396 KJ respectively, calculate the molar Heat of formation of ethane in KJ.

Solution

ΔH = $\Delta H_{\text{Formation}}$ product − $\Delta H_{\text{Formation}}$ reactant

First we find

ΔH_f product = Molar Heat of formation of H_2O + Molar Heat of formation of CO_2

2 Moles of CO_2 = 2 × -396 = -792
2 moles of H_2O = 2 × -286 = -572
∴ ΔH_f product = (-792) + (-572) = -1364 KJ

Substituting in the first equation

$\Delta H_{\text{reaction}}$ = $\Delta H_{\text{Formation}}$ product − $\Delta H_{\text{Formation}}$ reaction

-1428 = - 1364 − ΔH $_{Formation}$ ethene

∴ Δ H $_{Formation}$ Ethene = +1428 − 1364

∴ Δ H $_{Formation}$ Ethene = 64KJ

6. ½ N_2(g) + ½ O_2(g) → NO(g) ΔH° = 89KJ mol^{-1},
If the entropy change for the reaction above at 25°C is 11.8 J mol^{-1}, calculate the change in free energy, ΔG° for the reaction at 25°C.

Solution

Using, ΔG   =   ΔH − TΔS
ΔH   =   89 KJ mol^{-1} = 89,000J
T   =   25 + 273 = 298K
ΔS   =   11.8J mol^{-1}

Substituting

ΔG   =   89,000 − 298 x 11.8
ΔG   =   89,000 − 35164
ΔG   =   85483.6J
ΔG   =   85.48.KJ

7. How much heat will be liberated if 8g of hydrogen burn in excess oxygen according to the following equation.

2H_2(g) + O_2(g) → 2H_2O(g), ΔH = - 571.7 KJ [H = 1]

Solution

From the equation the combustion of 1 mole of hydrogen yields -571.7KJ of energy that is 2g of Hydrogen yields -571.7KJ of energy.

∴ 8g of Hydrogen will yield xKJ of energy

Solve for x,

x = 8 x -571.7
      2   =   -2286.8KJ

∴ When 8g of hydrogen is burnt in excess oxygen, -2286.8KJ of energy will be liberated.

8. Calculate the rise in temperature when 25cm^3 of 0.1M sodium hydroxide and 22cm^3 of 1.5M hydrochloric acid are mixed, starting from the same temperature. [Specific heat capacity of all solutions are 4.2 KJ/kg, heat of neutralization 15 − 57KJ mol^{-1}).

Solution

First we find the number of mole in 25cm^3 of 0.1M NaOH solution.

1000cm^3 contains 0.1 mole of NaoH

∴ 25cm^3 will contain x mol of NaOH

x   =   2.5 x 0.1
              1000

    =   0.0025 mole of NaOH

Total volume of mixture   =   25 x 22 = 47cm^3

Assuming the solution has the same density as water.

Mass of solution = 0.047kg

Applying

Heat gained = mass × specific.heat.capacity × rise in temp

Heat of neutralization of 0.0025 mol of NaOH

= −51 × 0.0025 = −0.1425KJ

Substituting in the above equation

− 0.1425 KJ = 0.047 × 4.2 × Δt

∴ Rise in temp = $\dfrac{0.1425}{0.1974}$

= 0.72°C

9. The standard electrode potentials for Pb/Pb^{2+} and Cu/Cu^{2+} are − 0.13v and +0.34v respectively at 293k. Calculate the free energy change when a lead rod is dipped into a lead (ii) tetraoxosulplate (vi) solution and a copper rod is dipped into a copper (ii) tetraoxosulphate (vi) solution. If both cells are connected together by a salt-bridge.

If the heat of the reaction

$Pb(s) + Cu^{2+}(aq) \rightarrow Pb^{2+}(aq) + Cu(s)$ is -11.2 KJ mol^{-1}

Calculate the entropy change, TΔS.

E^O = +0.34 − (−0.13)

= 0.47v at 293 K

Solution

$$\Delta G^\circ = -nFE^\circ$$
$$= -2 \times 96500 \times 0.47$$
$$= -90.71 \text{ KJ mol}^{-1}$$
$$\Delta H = -112 \text{ KJmol}^{-1}$$

Applying, $\Delta G = \Delta H - T\Delta S$

$-90.71 = -112 - 293 \Delta S$

$21.29 = -293\Delta S$

$\Delta S = \dfrac{21.29}{-293}$

$\Delta S = -0.073 \text{ KJK}^{-1}$
$\quad = -72.6 \text{ JK}^{-1}$

QUESTIONS

1. $Fe_2O_3 (s) + 2Al (s) \rightarrow Al_2O_3 (s) + 2fe(s)$

If the heats of formation of Al_2O_3 and Fe_2O_3 are -1670 KJ mol^{-1} and -822 KJ Mol^{-1} respectively, the enthalpy change in KJ for the reaction is?

2. $2SO_2 (g) + O_2 (g) \rightarrow 2SO_3 (g)$.

In the reaction above, the standard heats of formation of SO_2 (g) and SO_3 (g) are – 297 KJ mol^{-1} and – 396 KJ mol^{-1} respectively. The heat change of the reaction is ?

3. 3.25dm^3 of a gas at a pressure of 7.25 atm and a temperature of 37.8°C was allowed to expand against an external pressure of 1 atm. The temperature dropped to 24.5°C. If the heat lost was 2.16KJ. Calculate the change in the internal energy of the gas assuming the gas behaves ideally.

4. Ammonia gas prepared by the Haber process is very soluble in water to form ammonium hydroxide. Assuming the reaction occurs as follows,
$NH_3(g) + H_2O(l) \rightarrow NH_4OH(aq)$
Calculate the heat of reaction given the following heat of formation.
NH3(g) = -146.19KJ, NH_4OH = -366.7KJ, $H_2O(l)$ = -285.8KJ

5. The heat of combustion of ethanol, C_2H_5OH according to the equation
$$C_2H_5OH(l) + 3O_2(g) \rightarrow 2CO_2(g) + 3H_2O(g)$$
is – 1235KJ. Calculate the heat of formation of C_2H_5OH (l) given $\Delta H_f^o CO_2$ = -393.5KJ and ΔH_f^o $H_2O(l)$ = -241.8KJ

RATE LAWS

The rate of a reaction can be expressed as a function of concentration.

For a reaction

$aA + bB \rightarrow cC + dD$

The rate of reaction is related to the concentration of the reactants/products by

Rate = $k[A]^x [B]^y$

Rate = $k[C]^m [D]^n$

These equations, which relate the rate to the concentrations, are called the rate law or rate equation. K is the specific rate constant x, y, m and n are called the order of reaction. The sum of the exponents in each equation (x +y or m+n) gives the overall reaction order. The order of a reaction could be zero, first, second, third or higher. The exponents in the rate equation could be determined by measuring the initial rate of reaction for various initial concentrations of reactions.

For instance

In the reaction between H_2 and I_2 the following initial rates were obtained

$H_2(g) + I_2(g) \rightarrow 2HI(g)$

Experiment	Initial concentration, mol^{-1}		Initial reaction rate, Mol^{-1}S^{-1}
	[H_2]	[I_2]	
1	0.031	0.054	4.920×10^{-4}
2	0.093	0.054	1.476×10^{-3}
3	0.93	0.018	4.929×10^{-4}

Use the data to determine the overall order of the reaction.

Solution

The rate equation for the reaction is
Rate = K [H_2]^X [I_2]^y

Using the various initial rates we find the values of the exponents x and y in the rate equation. Let R_1 and R_2 represent rates for experiment 1 and 2 respectively.

$$\frac{R_2}{R_1} = \frac{k(0.093)^X (0.054)^y}{k(0.031)^X(0.054)^y} = \frac{1.476 \times 10^{-3}}{4.920 \times 10^{-4}} = 3$$

$$\frac{10.093)^X}{(0.031)^X} = 3$$

x = 1

With respect to H_2 the reaction is first order

Also

$$\frac{R_2}{R_1} = \frac{k(0.093)^X (0.054)^y}{k(0.093)^X(0.018)^y} = \frac{1.476 \times 10^{-3}}{4.92 \times 10^{-4}} = 3$$

$3^y = 3$

y = 1

Also the reaction is first order with respect to I_2

Overall order of reaction = x + y = 1+ 1 = 2

Zero Order Reactions

These are reactions, which occur at a rate that appears independent of the concentration of reactants. And they occur at a constant rate.

(Examples: photosynthesis, some enzyme catalyzed reactions).

Rate = $k[A]^0$ = Constant = K

The concentration $[A]_t$ of the reactant at time t is related to the initial concentration $[A]_o$ by $[A]_o = [A]_t + k_t$

A plot of concentration, [A], against time for a zero order reaction gives a straight line, with a negative slope (rate constant, k).

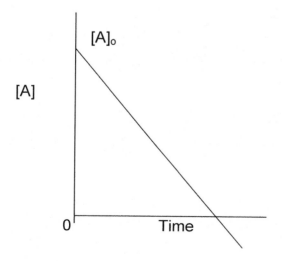

FIRST ORDER REACTION

For a first order reaction the sum of the exponents in the rate equation would be equal to 1.

For a first order reaction

A → Products

Rate = k[A] (the unit of rate constant is sec^{-1})

The concentration of the reactant $[A]_t$ at time, t is related to the initial concentration $[A]_o$ by

$$\text{Log } [A]_t = \frac{-kt}{2.303} + \text{Log } [A]_o$$

$$\text{Log } \frac{[A]_o}{[A]_t} = \frac{k_t}{2.303}$$

Examples

Decomposition of H_2O_2,

$2H_2O_2(l) \rightarrow 2H_2O(l) + O_2(g)$

Decomposition of N_2O_5

$2N_2O_5(g) \rightarrow 4NO_2(g) + O_2(g)$

A plot of log $[A]_t$ against time would give a straight line graph, with the

$$\text{slope} = \frac{-k}{2.303}$$

SECOND ORDER REACTIONS

For a second order reaction the sum of the components in the rate equation would be equal to 2, because it may involve one, or two reactants.

A → Products   or
A+B → Products
Rate = $k[A]^2$
or Rate = $k[A][B]$

The integrated rate equation is given as

$$\frac{1}{[A]_t} = \frac{1}{[A]_o} + k_t \quad \text{(for one reactant)}$$

If we plot $1/[A]_t$ against time we would have a straight line graph, with the slope equal to k

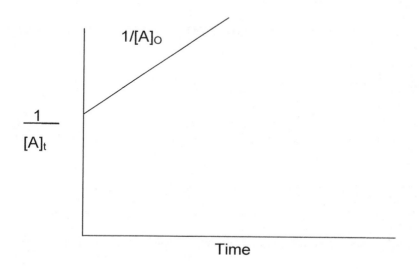

The intercept of the graph is equal to the reciprocal of the initial concentration.

HALF-LIFE CALCULATIONS

Half-life of a radioactive element is the time taken for half the nuclei in a given sample of a radioactive element to decay.

$$\log \frac{A}{A_O} = \frac{-Kt}{2.303}$$

Where A_O = amount of radioactive element present initially.

A = amount present after time, t

t = time

k = rate constant (delay constant)

Half Life, $t_{1/2}$

$$t_{1/2} = \frac{0.693}{k}$$

Exercises

1. The isotope $^{137}_{55}Cs$ has a half life of 30 years. What amount of a 0.001g sample of this isotope would remain after 16 years.

Solution

Using the half-life equation we find the decay constant.

$$t_{1/2} = \frac{0.693}{k}$$

$$\therefore k = \frac{0.693}{t_{1/2}} = \frac{0.693}{30} = 0.0231 \text{ yr}^{-1}$$

Then applying the rate law equation we calculate the fraction remaining at the end of 16 years.

$$\log \frac{A}{A_O} = \frac{kt}{2.303}$$

$$\log \frac{A}{A_O} = \frac{0.0231 \text{ yr}^{-1} \times 16 \text{yrs}}{2.303}$$

$$\log \frac{A}{A_O} = -0.161$$

$$\therefore \frac{A}{A_O} = 0.690$$

Amount remaining after 16 years
= 0.690 x 0.001g
= 0.00069g

Questions

1. Calculate the amount of 0.07og sample that is remaining at the end of 20 years for the isotope $^{90}_{38}Sr$, which has a half-life of 28.1 year and is known to decay by alpha emission.

2. Calculate the amount of 25g of ^{210}Po whose half-life is 140 days which remains after 35 days?

Equilibrium Constant

According to the Law of Mass Action, "At constant temperature, the rate of reaction is proportional to the active masses of each of the reactants". Let the active mass be the concentration of the substance raised to a power, which is equal to its numerical coefficient (for gases, the partial pressure is used instead of its concentration).
For a reaction like the one below,

$$xA + yB \xrightarrow{r} C$$

In which x moles of reactant A, reacts with y moles of reactant B, at a rate of reaction, r, to form product C.
From the law of mass action we can deduce that:

$r \; \alpha \; (\text{concentration of A})^x$

Also, $r \; \alpha \; (\text{concentration of B})^y$

It follows that, $r \; \alpha \; (\text{concentration of A})^x \times (\text{concentration of B})^y$
The square bracket, [ ] is used to represent the concentration in moles per dm^3 of a substance.
So, $r \; \alpha \; [A]^x [B]^y$

We can replace the variation by an equality sign. Introducing a constant, k in the process (as usual).
$r = K [A]^x [B]^y$
K is the velocity constant of that reaction at that temperature.
If we apply the same thing to a reversible reaction of the form

$$xA + yB \underset{r_2}{\overset{r_1}{\rightleftharpoons}} eC + gD$$

The rate of forward reaction, $r_f = K_f [A]^x.[B]^y$

The rate of the backward reaction, $r_b = K_b [C]^e.[D]^g$.

At equilibrium, that is when the rate of forward reaction is equal to the backward reaction.

$$r_f = r_b$$

That is, $K_f [A]^x.[B]^y = K_b [C]^e.[D]^g$

Therefore, $Kf/Kb = [C]^e.[D]^g / [A]^x.[B]^y = K$

K is the equilibrium constant for that reaction at that temperature.

SOLUBILITY PRODUCT CONSTANT

When substances dissolve in water, they dissociate into ions. An equilibrium is usually established when a saturated solution is formed, and the solution is in equilibrium with the undissolved solid. At equilibrium the solid state will continue to dissolve while the dissolved solute precipitates or crystallizes out of the solution, leading to the formation of more of the solid solutes, with the rate of these opposing processes being equal.

For instance, for a salt like silver chloride, AgCl, the equilibrium can be expressed as, $AgCl(s) \underset{crystallization}{\overset{dissolutions}{\rightleftharpoons}} Ag^+(aq) + Cl^-(aq)$

Applying the law of mass action, the equilibrium constant, K would be given as,

$$K = \frac{[Ag^+][Cl^-]}{[AgCl]}$$

The concentration of the solid is constant and can be incorporated into the equilibrium constant to give a new constant, the solubility product constant (solubility product, Ksp) (ion product)

$$K \times [AgCl](s) = [Ag^+] \times [Cl^-] = Ksp$$

The solubility product Ksp of a salt is equal to the product of the molar concentrations of the ions involved in the equilibrium each raised to the power of its stoichiometric coefficient (in the balanced equation).

Since, the solubility of a substance varies with temperature, the value of Ksp changes with temperature.

For the following equilibrium the solubility product (ion product) can be written as,

1. $BaSO_4 (s) \rightleftharpoons Ba^{2+} (aq) + SO_4^{2-} (aq)$

    $Ksp = [Ba^{2+}] \times [SO_4^{2-}]$

2. $PbCl_2 (s) \rightleftharpoons Pb^{2+} (aq) + 2Cl^- (aq)$

    $Ksp = [Pb^{2+}] \times [Cl^-]^2$

3. $MgNH_4PO_4 (s) \rightleftharpoons Mg^{2+} (aq) + NH^+ (aq) + PO_4^{3-} (aq)$

    $Ksp = [Mg^{2+}] \times [NH_4^+] \times [PO_4^{3-}]$

Calculations on solubility product

1. $0.250 dm^3$ of a saturated solution of AgCl at 25^oC contains 0.101g of AgCl. Calculate the solubility product constants for AgCl.

Solution

First, we write the equilibrium equation for the salt and its ions present in solution.

$$AgCl \rightleftharpoons Ag^+(aq) + Cl^- (aq)$$

This shows that for every mole of AgCl there is one mole of Ag^+ and also one mole of Cl^-.

We determine the molar solubility of AgCl

Since $0.25 dm^3$ of the saturated solution of AgCl contains 0.101g of AgCl.

∴ (The molar solution i.e.) $1 dm^3$, will contain Xg of AgCl

i.e. $0.25 dm^3$ → 0.101g of AgCl

$1.0 dm^3$ → Xg of AgCl

If we cross multiply

$0.25 \times X = 1.0 \times 0.101$

$X = \dfrac{1.0 \times 0.101}{0.25} = 0.404$

∴ The solubility of AgCl is 0.404g per dm^3 converting this to molar solubility that is to moles per dm^3.

Remember,

Concentration in gdm^{-3} = conc $moldm^{-3}$ × molar mass

∴ Molarity (mol dm^{-3}) = $\dfrac{\text{Conc in } gdm^{-3}}{\text{Molar Mass}}$

Molar mass of AgCl = 108 + 35.5 = 143.5

Molarity = $\dfrac{0.404}{143.5}$ = 0.0028 $moldm^{-3}$

The solubility of AgCl = 0.0028 mol dm^{-3}

= 2.8×10^{-3} mol dm^{-3}

This could be related to the concentrations of the ions in solution through the stoichiometric coefficients in the balanced equilibrium equations.

The concentration of Ag^+

$[Ag^+]$ = $\dfrac{2.8 \times 10^{-3} \text{ mol AgCl} \times 1\text{mol }[Ag^+] \text{ per 1mol AgCl}}{dm^3}$

= 2.8×10^{-3} mol Ag^+ per dm^3

The concentration of Cl^-,

$[Cl^-] = 2.8 \times 10^{-3}$ moldm^3 AgCl $\times$ 1 mol(Cl^-) per 1 mol AgCl

$\qquad\qquad\qquad = \quad 2.8 \times 10^{-3}$ mol Cl^- dm^{-3}

$\qquad$ Ksp $\quad = \quad [Ag^+][Cl^-]$

$\qquad\qquad\qquad = \quad (2.8 \times 10^{-3})(2.8 \times 10^{-3})$

$\qquad$ Ksp $\quad = \quad 7.84 \times 10^{-6}$

2. The solubility of PbI_2 at $30^\circ C$ was found to be 1.21×10^{-3} moldm^{-3}. Calculate the solubility product constant for PbI_2.

Solution

First, we write the equilibrium equation for the salt and its ion present in solution.

$\qquad PbI_2 \text{ (s)} \rightleftharpoons Pb^{2+}(aq) + 2I^-(aq)$

This shows that for every mole of PbI_2 there are two moles of I^- and one mole of Pb^{2+} ions.

The concentration of Pb^+,

$[Pb^{2+}] = 1.21 \times 10^{-3}$ mol PbI_2 per dm^3 × 1 mol (Pb^{2+}) per 1 mol PbI_2
$\qquad = 1.21 \times 10^{-3}$ mol Pb^{2+} per dm^3

The concentration of I^-

$[I^-] = 1.21 \times 10^{-3}$ mol PbI_2 per dm^3 × 2 mol $[I^-]$ per 1 mol PbI_2
$\qquad = 2.42 \times 10^{-3}$ mol I^- per dm^{-3}

$\text{Ksp} = [Pb^{2+}] \times [I^-]^2$
$\qquad = (1.21 \times 10^{-3}) \times (2.42 \times 10^{-3})^2$
$\qquad = 7.09 \times 10^{-9}$

3. 200ml sample of saturated solution of $Cr(OH)_3$ at $25^\circ C$ contains 0.027g of the compound. Calculate the solubility product of $Cr(OH)_3$.

Solution

Equilibrium equation

$$Cr(OH)_3 \text{ (s)} \rightleftharpoons Cr^{3+} \text{ (aq)} + 3\, OH^- \text{ (aq)}$$

From this equation, the stoichometric equivalence is
1 mol $Cr(OH)_3$ ≡ 1 mol Cr^{3+} ≡ 3 mol OH^-

To find the molar solubility of $Cr(OH)_3$ since 200ml (cm^3) of saturated $Cr(OH)_3$ solution contains 0.0227g of $Cr(OH)_3$
1000ml (cm^3) of the saturated solution will contain Xg of $Cr(OH)_3$
as usual cross multiplying
200 × X   =   1000 × 0.0227

$$X = \frac{1000 \times 0.0227}{200}$$

$X = 0.1135$g of $Cr(OH)_3$ Per dm^3 converting this to $moldm^{-3}$

X in $Moldm^{-3}$ = Mass/Molar Mass

$$= 0.1135/103 = 0.001102 \text{ moldm}^{-3}$$

Solubility of $Cr(OH)_3 = 1.102 \times 10^{-3}$ moldm^3

The Concentration of $Cr^{3+}, [Cr^{3+}]$

$= 1.102 \times 10^{-3}$ mol of $Cr(OH)_3$ per dm^3 x 1 mol Cr^{3+} per 1 mol $Cr(OH)_3$

$$= 1.102 \times 10^{-3} \text{ moldm}^{-3} \text{ of } Cr^{3+}$$

The concentration of $OH^-, [OH^-]$

$= 1.102 \times 10^{-3}$ mol of $Cr(OH)_3$ perdm^3 x 3 molCr^{3+} per 1 mol $Cr(OH)_3$

$$= 3.31 \times 10^{-3} \text{ mol dm}^{-3} \text{ of } OH^-$$

$KSP = [Cr^{3+}] \times [OH^-]^3 = (1.102 \times 10^{-3}) \times (3.31 \times 10^{-3})^3$

$$= 39.96 \times 10^{-12}$$
$$= 3.996 \times 10^{-11}$$

4. Calculate the molar solubility of PbI_2 in water given that the solubility product constant is 1.7×10^{-6}.

Solution

The equilibrium equation is

$$PbI_2 \text{ (s)} \rightleftharpoons Pb^{2+} \text{ (aq)} + 2I^- \text{ (aq)}$$

The stoichometric equivalence

$$1 \text{ mol } PbI_2 = 1 \text{ mol } Pb^{2+} = 2 \text{ mol } I^-$$

Let us assume that the molar solubility of PbI_2 = n, that is the number of moles of PbI_2 that is dissolved in $1dm^3$ of the solution

The concentration of the ions becomes

$[Pb^{2+}]$ = n
$[I^-]$ = 2n
K_{sp} = 1.7×10^{-6}
K_{sp} = $[Pb^{2+}] \times [I^-]^2$

Substituting the values (of concentration) in the above equation, we get

$$1.7 \times 10^{-6} = n \times (2n)^2$$
$$1.7 \times 10^{-6} = n \times 4n^2$$
$$1.7 \times 10^{-6} = 4n^3$$

$$\frac{1.7 \times 10^{-6}}{4} = \frac{4 \times n^3}{4}$$

$$0.425 \times 10^{-6} = n^3$$
$$n = \sqrt[3]{0.425 \times 10^{-6}}$$
$$= 0.00752$$
$$= 7.52 \times 10^{-3} \text{ moldm}^{-3}$$

The molar solubility of PbI_2 = 7.52 $moldm^{-3}$

5. Calculate the molar solubility of $PbSO_4$, given that the solubility product constant of $PbSO_4$, is 2×10^{-8}

Solution

The equilibrium equation is

$$PbSO_4 (s) \rightleftharpoons Pb^{2+}(aq) + SO_4^{2-}(aq)$$

The stoichiometric equivalence

1 mol of $PbSO_4$ ≡ 1 mol Pb^{2+} = 1 mol SO_4^{2-}

Let the molar solubility of $PbSO_4$ = n

$[Pb^{2+}]$ = n

$[SO_4^{2-}]$ = n

Ksp = $[Pb^{2+}] \times [SO_4^{2-}]$

Substituting the values

2×10^{-8}	=	n x n
2×10^{-8}	=	n^2
n	=	$\sqrt{2 \times 10^{-8}}$
n	=	0.00014
	=	1.4×10^{-4} $moldm^{-3}$

The molar solubility of $PbSO_4$ = 1.4×10^{-4} $moldm^{-3}$

6. Calculate the molar solubility of $BaSO_4$, given that the solubility product constant of $BaSO_4$ is 1.5×10^{-9}

Solution

The equilibrium equation is

$$BaSO_4(s) \rightleftharpoons Ba^{2+}(aq) + SO_4^{2-}(aq)$$

The stoichiometric equivalence is

1 mol of $BaSO_4$ ≡ 1 mol of Ba^{2+} ≡ 1 mol of SO_4^{2-}

Let the molar solubility of $BaSO_4$ be n

$[Ba^{2+}]$ = n

$[SO_4^{2-}]$ = n

K_{sp} = $[Ba^{2+}] \times [SO_4^{2-}]$

Substituting the values

1.5×10^{-9} = n x n

n^2 = 1.5×10^{-9}

n = $\sqrt{1.5 \times 10^{-9}}$

n = 3.87×10^{-5}

HYDROGEN ION CONENTRATION

Pure water ionizes slightly to yield equal amount of hydrogen ions, H^+ and hydroxyl ion, OH^-.

The concentration of Hydrogen ions, $[H^+]$ and hydroxyl ion $[OH^-]$ is equal to 10^{-7} moldm^{-3}

$[H^+] = [OH^-] = 10^{-7}$ moldm^{-3}

The product of these ionic concentrations gives the ionic product of water, kw

$Kw = [H^+][OH^-]\ 10^{-7} \times 10^{-7} = 10^{-14}$ moldm^{-3}

pH

The pH of a solution is the negative logarithm of the hydrogen ion concentration to the base 10.

Hence for a solution whose hydrogen ion concentration is 10^{-6} moldm^{-3}, the pH value that is the degree of acidity or alkalinity, could be determined as follows:

Remember

pH = $Log_{10}[H^+]$

and $[H^+]$ = 10^{-6}

$Log_{10}[H^+]$ = $Log_{10}10^{-6}$ = -6 (from law of logarithms $Log_{10}10^{-x} = x$)

pH = $-Log_{10}[H^+]$ = -(-6) = 6

∴ pH value of solution = 6

We could determine both the pH value and the pOH value from the hydrogen ion concentration of a solution since the product of the hydrogen ion concentration and hydroxyl ion concentration for a solution is always constant (and equals 10^{-14} moldm^{-3}).

$[H^+][OH^-] = 10^{-14}$ moldm^{-3}

∴   pH + pOH = 14

pOH = 14 - pH

pOH = hydroxide ion index

Exercises

1. Calculate the hydrogen and hydroxide ion concentrations for the following solutions
(i)   0.01M Hydrochloric acid
(ii)   0.02M tetraoxosulphate (vi) acid
(iii)   0.001M sodium hydroxide solution

Solution
(i)   The ionization equation for HCl acid is   HCl (aq) → H^+(aq) + Cl^- (aq)

From the equation, one mole of hydrochloric acid yields one mole of hydrogen ion on ionization.

So a 0.01M (0.01 moldm^{-3}) solution would yield

1 x 0.01 moldm^{-3} of hydrogen ion

∴ [H$^+$] = 1 x 0.01 moldm^{-3} = 0.01 = 10^{-2} moldm^{-3}

Applying, [H$^+$][OH$^-$] = 10^{-14}

$$10^{-2} \times [OH^-] = 10^{-14}$$

$$[OH^-] = \frac{10^{-14}}{10^{-2}} = 10^{-12} \text{ moldm}^{-3}$$

(ii) The ionization equation is

H$_2$SO$_4$ (aq) → 2H$^+$(aq) + SO$_4^{2-}$(aq)

From the equation one mole of tetraoxosulphate (vi) acid yields two moles of hydrogen ion on ionisation

So a 0.01M solution will yield 2 x 0.01 moldm^{-3} of hydrogen ion

∴ [H$^+$] = 2 x 0.01 = 0.02 = 2 x 10^{-2} moldm^{-3}

Applying, [H$^+$][OH$^-$] = 10^{-14}

$$2 \times 10^{-2} \times [OH^-] = 10^{-14}$$

$$[OH^-] = \frac{10^{-14}}{2 \times 10^{-2}} = 0.5 \times 10^{-12}$$

$$= 5 \times 10^{-13} \text{ moldm}^{-3}$$

(iii) The ionization equation is

NaOH (aq) → Na$^+$(aq) + OH$^-$(aq)

From the equation one mole of sodium hydroxide yields one mole of hydroxyl ion on ionization

So a 0.001 (moldm^{-3}) solution will yield

1 x 0.001 moldm^{-3} of hydroxyl ion

∴ $[OH^-] = 1 \times 0.001 = 0.001 = 10^{-3}$ moldm^{-3}

Applying, $[H^+][OH^-] = 10^{-14}$

$[OH^-] \times 10^{-3} = 10^{-14}$

$[H^+] = \dfrac{10^{-14}}{10^{-3}} = 10^{-11}$ moldm^{-3}

2. Calculate the pH of the following solutions, with hydrogen ion concentration

(i) 1×10^{-5} moldm^{-3}
(ii) 2×10^{-4} moldm^{-3}
(iii) 5×10^{-8} moldm^{-3}

Solution

(i) $[H^+] = 1 \times 10^{-5} = 10^{-5}$

Applying, pH = $-Log_{10}[H^+]$

$Log_{10}[H^+] = Log_{10}10^{-5} = -5$

pH = $-Log_{10}[H^+] = -(-5) = 5$

pH = 5

(ii) $[H^+] = 2 \times 10^{-4}$ moldm^{-3}

When faced with a problem like this, find the Logarithm of 2, (log 2 = 0.30)

So, $[H^+] = 2 \times 10^{-4}$ becomes

$= 10^{0.30} \times 10^{-4}$

$= 10^{-3.70}$

Now, $\log [H^+] = \log 10^{-3.70} = -3.70$

pH $= -\log [H^+] = -(-3.70) = 3.70$

pH $= 3.70$

(iii) $[H^+] = 5 \times 10^{-8}$ moldm^{-3}

$\log 5 = 0.70$

$[H^+] = 10^{-0.70} \times 10^{-8} = 10^{-7.30}$

$\log [H^+] = \log 10^{-7.30} = -7.30$

pH $= -\log[H^+] = -(-7.30) = 7.30$

pH $= 7.30$

3. Find the hydrogen ion concentration for a solution with a pH of 3.4.

Solution

Remember,

pH $= -\log_{10}[H^+]$

Since, pH = 3.4

We have $3.4 = -\log_{10}[H^+]$

$\therefore \log_{10}[H^+] = -3.4$

To find the value of $[H^+]$, we have to get the antilog of -3.4, but first

$-3.4 = -4.0 \times 0.6 = 4.6$

$\therefore [H^+] = $ antilog 4.6

$= 4 \times 10^{-4}$ moldm^{-3}

Hydrogen ion concentration $= 4 \times 10^{-4}$ moldm^{-3}

4. The POH of a solution of 0.25 moldm^{-3} of hydrochloric acid is?

Solution

Equation of ionization is

$$HCl\ (aq) \rightarrow H^+(aq) + Cl^-\ (aq)$$

From the equation one mole of hydrochloric acid yields one mole of hydrogen ion on ionization.

So a 0.25 moldm^{-3} solution yields 1 x 0.25 moldm^{-3} of hydrogen ion

$[H^+]$ = 1 x 0.25 = 0.25 = 2.5 x 10^{-1} moldm^{-3}

Remember, pH + pOH = 14

But first we have to find the value of pH

Applying, pH = -Log$_{10}$[H$^+$]

(Remember log 2.5 = 0.40)

$[H^+]$ = 10$^{-0.40}$ x 10^{-1} = 10$^{-0.60}$

Log [H$^+$] = Log 10$^{-0.60}$ = - 0.60

pH = - Log$_{10}$[H$^+$] = - (-0.60) = 0.60

pH = 0.60, substituting in

pH + pOH = 14

0.60 + POH = 14

pOH = 14 – 0.60

= 13.40

pOH = 13.40

5. 1.0dm³ of distilled water was used to wash 2.0g of a precipitate of AgCl. If the solubility product of AgCl is 2.0×10^{-10} moldm^{-3}. What quantity of silver was lost in the process?

Solution

The quantity of AgCl lost would be the amount of AgCl, which could dissolve, and thus is the same thing as the solubility of AgCl. We could calculate the solubility of AgCl from its solubility product.

Remember,
$$K_{sp} = [Ag^+] \times [Cl^-]$$

From the ionization equation of AgCl

$$AgCl\ (s) \rightleftharpoons Ag^+(aq) + Cl^-(aq)$$

Let the solubility of AgCl be n
Substituting in

$$K_{sp} = [Ag^+][Cl^-]$$
$$2.0 \times 10^{-10} = n \times n$$
$$n^2 = 2.0 \times 10^{-10}$$
$$n = \sqrt{2.0 \times 10^{-10}}$$
$$n = 1.4 \times 10^{-5}$$

The solubility of AgCl is 1.4×10^{-5} moldm^{-3}
∴ The quantity of AgCl lost was 1.4×10^{-5} moldm^{-3}

6. The solubility product of $Cu(Ol_3)_2$ is 1.08×10^{-7}. Assuming that neither ion reacts appreciable with water to form H^+ and OH^- what is the solubility of this salt.

Solution

First we write the equilibrium equation for the salt.

$$Cu(Ol_3)_2 \text{ (s)} \rightleftharpoons Cu^{2+}\text{(aq)} + 2Ol_3^-\text{ (aq)}$$

The stoichiometric equivalence

1 mol $Cu(Ol_3)_2$ ≡ 1 mol Cu^{2+} ≡ 2 mol Ol_3^-

Let us assume the molar solubility of $Cu(Ol_3)_2$ to be n

Hence,

$[Cu^{2+}]$ = n

$[Ol_3^-]$ = 2n

Applying, $Ksp = [Cu^{2+}][Ol_3^-]^2$

$1.08 \times 10^{-7} = n \times (2n)^2$

$1.08 \times 10^{-7} = n \times 4n^2$

$4n^3 = 1.08 \times 10^{-7}$

$\dfrac{4n^3}{4} = \dfrac{1.08 \times 10^{-7}}{4}$

$n^3 = 2.7 \times 10^{-8}$

$n = \sqrt[3]{2.7 \times 10^{-8}}$

$n = 0.003$

The molar solubility of $Cu(OI_3)_2$ is $0.003\ moldm^{-3}$ or $3.0 \times 10^{-3}\ moldm^{-3}$.

7. A saturated solution of AgCl was found to have a concentration of
$1.30 \times 10^{-5}\ moldm^{-3}$. The solubility product of AgCl therefore is?

Solution

First we write the equilibrium equation for AgCl

$AgCl \rightleftharpoons Ag^+(aq) + Cl^-(aq)$

The stoichiometric equivalence
1 mol AgCl ≡ 1 mol Ag^+ ≡ 1 mol Cl^-

The molar solubility of AgCl is the same as the concentration of its saturated solution which is $1.30 \times 10^{-5}\ moldm^{-3}$

$[Ag^+] = 1 \times 1.30 \times 10^{-5}\ moldm^{-3} = 1.30 \times 10^{-5}\ moldm^{-3}$

$[Cl^-] = 1 \times 1.30 \times 10^{-5}\ moldm^{-3} = 1.30 \times 10^{-5}\ moldm^{-3}$

Applying, $Ksp = [Ag^+][Cl^-]$

> **Remember** Ksp = Solubility Product

$Ksp = 1.30 \times 10^{-5} \times 1.30 \times 10^{-5}$
$= 1.69 \times 10^{-10}$

The solubility product of AgCl = 1.69×10^{-10}

QUESTIONS

1. What is the concentration of H^+ ions in moles per dm^3 of a solution of pH 4.398?

2. Calculate the pH of the following solutions
(i)   0.001M HCl
(ii)   3.2 x 10^{-3}M NaOH
(iii)   A solution containing 2.28g of KOH per dm^3

Practice Questions

1. Given that 6.10g of a metal, M (RMM = 27) reacts completely with 22.9g of chlorine (RMM = 35.5) to give 29.0g of a metallic chloride, what is the empirical formula of the chloride

2. 30g of potassium trioxochlorate (v) was heated to a constant mass. What mass of oxygen was produced? And calculated the volume of oxygen produced at 800mm Hg and 45°C.
[O = 16, Cl = 35.5, K = 39]

3. 2g of carbon (iv) oxide is bubbled into excess lime water. What mass of the precipitate is formed?

4. 30g of impure calcium trioxocarbonate (iv) reacts with excess hydrochloric acid, to liberate 3.38dm^3 of carbon(iv) oxide at s.t.p., calculate the percentage purity of the impure salt.

5. Find the volume of oxygen at s.t.p. required for the complete combustion of 30cm^3 of ethane? [molar volume = 22.4dm^3 at s.t.p]

6. Calculate the volume of oxygen at s.t.p. that will be produced by heating 300g of mercury (ii) oxide?
[HgO = 217, molar volume of gases = 22.4dm^3 at s.l.p]

7. Naturally occurring boron is made up of 20% by mass of the $^{10}_{5}B$ isotope and 80% by mass of the $^{11}_{5}B$ isotope. Calculate the relative atomic mass of boron

8. What volume of air is required for the complete combustion of 5g of butane at s.t.p. If given that oxygen is 21% by volume of air.

9. A given volumes of hydrogen gas weighed 3g and an equal volume of a gas X weighs 11g. Find the relative molecular mass of the gas, X.

10. Find the percentage by mass of water of crystallization in FeSO$_4$ 7H$_2$O?
(Fe = 56, S = 32, O = 16, H = 1)

11. A certain sample of seawater at 75°C contains 47g of potassium chloride and 23g of sodium chloride per 100g. At 11°C the sample was found to contain 35g of potassium chloride and 40g of sodium chloride. Calculate the amount of potassium chloride that will be deposited if 60g of the seawater is cooled from 75°C to 11°C.

12. A Copper, zinc and silver voltammeter are connected in series. If a current of 10 amperes flows for 2 hours 15 seconds, calculate the masses of copper, zinc and silver deposited in each cell.
[Cu = s63.5, Zn = 65, Ag = 108].

13. Determine the volume of oxygen at 800mmHg and $15^{\circ}C$ liberated, when a copper (II) tetraoxosulphate (vi) solution is electrolysed between platinum electrodes, if 3.7g of copper is discharged at the cathode.

14. Calculate the oxidation number of iodine in KIO_3

15. Calculate the mass of copper which would be discharged by the same quantity of electricity which liberates $23dm^3$ of hydrogen at s.t.p
(Cu = 635, A = 1, molar volume of gases = $12.edm^3$ at s.t.p.)

16. 6.6g of a mixture of sodium trioxocarbonate (iv) and sodium hydrogen trioxocarbonate (iv) were heated to a constant mass of 4.2g. Calculate the percentage by mass of sodium trioxocarbonate (iv) in the mixture.
[C = 12, Na = 23, O = 16, H = 1].

17. 25cm³ of a hydrocarbon requires 75cm³ of oxygen for complete combustion to yield 50cm³ of carbon (iv) oxide. Find the molecular formula of the hydrocarbon.

18. A certain hydrocarbon has a vapour density of 39. Find its molecular formula if it is made up of 7.69% hydrogen by mass.

Relative Atomic Masses of Elements

Element	Symbol	Atomic Number	Relative Atomic Mass
Actinium	Ac	89	227
Aluminium	Al	13	26.98
Americium	Am	95	243
Antimony	Sb	51	121.75
Argon	Ar	18	39.95
Arsenic	As	33	74.92
Astatine	At	85	210
Barium	Ba	56	137.34
Berkelium	Bk	97	249
Beryllium	Be	4	9.01
Bismuth	Bi	83	208.98
Boron	B	5	10.81
Bromine	Br	35	79.904
Cadmium	Cd	48	112.40
Caesium	Cs	55	132.905
Calcium	Ca	20	40.08
Californium	Cf	98	251
Carbon	C	6	12.01
Cerium	Ce	58	140.12
Chlorine	Cl	17	35.45
Chromium	Cr	24	51.996
Cobalt	Co	27	58.933
Copper	Cu	29	63.546
Curium	Cm	96	247
Dysprosium	Dy	66	162.50
Einsteinium	Es	99	254

Relative Atomic Masses of Elements

Element	Symbol	Atomic Number	Relative Atomic Mass
Erbium	Er	68	167.26
Europium	Eu	63	151.96
Fermium	Fm	100	257
Fluorine	F	9	18.998
Francium	Fr	87	223
Gadolinium	Gd	64	157.25
Gallium	Ga	31	69.72
Germanium	Ge	32	72.59
Gold	Au	79	196.966
Hafnium	Hf	72	178.49
Hahnium	Ha	105	262
Helium	He	2	4.002
Holmium	Ho	67	164.93
Hydrogen	H	1	1.008
Indium	In	49	114.82
Iodine	I	53	126.904
Iridium	Ir	77	192.22
Iron	Fe	26	55.847
Krypton	Kr	36	83.80
Lanthanum	La	57	138.905
Lawrencium	Lw/Lr	103	260
Lead	Pb	82	207.19
Lithium	Li	3	6.939
Lutetium	Lu	71	174.97

Relative Atomic Masses of Elements

Element	Symbol	Atomic Number	Relative Atomic Mass
Magnesium	Mg	12	24.312
Manganese	Mn	25	54.938
Mendelevium	Md	101	258
Mercury	Hg	80	200.59
Molybdenum	Mo	42	95.94
Neodymium	Nd	60	144.24
Neon	Ne	10	20.183
Neptunium	Np	93	237
Nickel	Ni	28	58.71
Niobium	Nb	41	92.906
Nitrogen	N	7	14.006
Nobelium	No	102	259
Osmium	Os	76	190.2
Oxygen	O	8	15.999
Palladium	Pd	46	106.4
Phosphorus	P	15	30.973
Platinum	Pt	78	195.09
Plutonium	Pu	94	244
Polonium	Po	84	210
Potassium	K	19	39.102
Praseodymium	Pr	59	140.907
Promethium	Pm	61	145
Protactinium	Pa	91	231.035
Radium	Ra	88	226.025
Radon	Rn	86	222
Rhenium	Re	75	186.2
Rhodium	Rh	45	102.905

Element	Symbol	Atomic Number	Relative Atomic Mass
Rubidium	Rb	37	85.467
Ruthenium	Ru	44	101.07
Rutherfordium	Rf	104	261
Samarium	Sm	62	150.4
Scandium	Sc	21	44.955
Selenium	Se	34	78.96
Silicon	Si	14	28.086
Silver	Ag	47	107.868
Sodium	Na	11	22.989
Strontium	Sr	38	87.62
Sulphur	S	16	32.06
Tantalum	Ta	73	180.947
Technetium	Tc	43	99
Tellurium	Te	52	127.60
Terbium	Tb	65	158.925
Thallium	Tl	81	204.37
Thorium	Th	90	232.038
Thulium	Tm	69	168.934
Tin	Sn	50	118.69
Titanium	Ti	22	47.90
Tungsten	W	74	183.85
Uranium	U	92	238.029
Vanadium	V	23	50.941
Xenon	Xe	54	131.30
Ytterbium	Yb	70	173.04
Yttrium	Y	39	88.905
Zinc	Zn	30	65.37
Zirconium	Zr	40	91.22

Printed in Great Britain
by Amazon